Bug Life

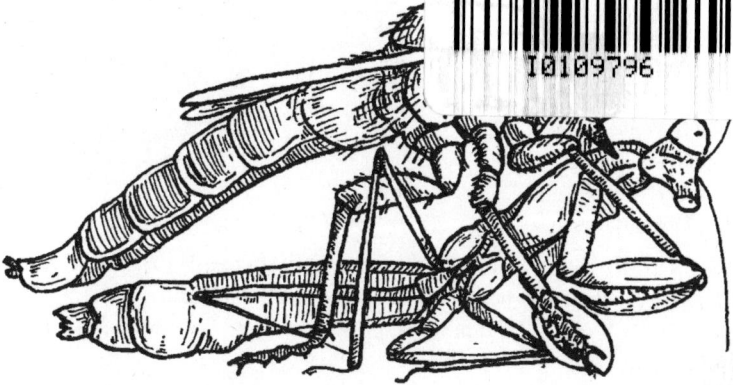

How Bees, Butterflies, and Other Insects Rule the World

Karyn Light-Gibson

Microcosm Publishing
Portland, OR | Cleveland, OH

Bug Life: How Bees, Butterflies, and Other Insects Rule the World
© Karyn Light-Gibson, 2025
This edition © Microcosm Publishing, 2025
First Edition, 3,000 copies, first published March 2025
ISBN 9781648413094
This is Microcosm #889
Designed by Joe Biel
Edited by Kandi Zeller

For a catalog, write or visit:

Microcosm Publishing
2752 N Williams Ave.
Portland, OR 97227
(503)799-2698
Microcosm.Pub

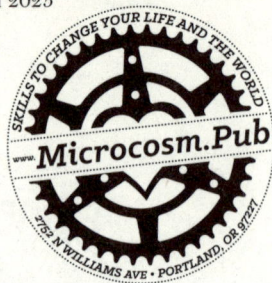

All the news that's fit to print at www.MicrocosmPub/Newsletter
Get more copies of this book at www.Microcosm.Pub/BugLife
Find more work by Karyn Light-Gibson at www.Microcosm.Pub/KarynLightGibson
Did you know that you can buy our books directly from us at sliding scale rates? Support a small, independent publisher and pay less than Amazon's price at **www.Microcosm.Pub.**

EU Safety Information: https://microcosmpublishing.com/gpsr

To join the ranks of high-class stores that feature Microcosm titles, talk to your rep: In the U.S. **COMO** (Atlantic), **ABRAHAM** (Midwest), **BOB BARNETT** (Texas, Oklahoma, Arkansas, Louisiana), **IMPRINT** (Pacific), **TURNAROUND** (UK), **UTP/MANDA** (Canada), **NEWSOUTH** (Australia/New Zealand), **Observatoire** (Africa, Europe), **IPR** (Middle East), **Yvonne Chau** (Southeast Asia), **HarperCollins** (India), **Everest/B.K. Agency** (China), **Tim Burland** (Japan/Korea), and **FAIRE** in the gift trade.

Global labor conditions are bad, and our roots in industrial Cleveland in the 70s and 80s made us appreciate the need to treat workers right. Therefore, our books are MADE IN THE USA and printed on post-consumer paper.

Library of Congress Control Number: 2025001406

MICROCOSM·PUBLISHING

MICROCOSM PUBLISHING is Portland's most diversified publishing house and distributor, with a focus on the colorful, authentic, and empowering. Our books and zines have put your power in your hands since 1996, equipping readers to make positive changes in their lives and in the world around them. Microcosm emphasizes skill-building, showing hidden histories, and fostering creativity through challenging conventional publishing wisdom with books and bookettes about DIY skills, food, bicycling, gender, self-care, and social justice. What was once a distro and record label started by Joe Biel in a drafty bedroom was determined to be *Publishers Weekly*'s fastest-growing publisher of 2022 and #3 in 2023, and is now among the oldest independent publishing houses in Portland, OR, and Cleveland, OH. Biel is also the winner of PubWest's Innovator Award in 2024. We are a politically moderate, centrist publisher in a world that has inched to the right for the past 80 years.

Contents

Introduction: Here's Why Bugs Are Cool

*H*ow do you feel about bugs, curious reader? Scared, apprehensive, fascinated? Many people feel all of those things and more when they see something scurrying around on six (or more) legs. Even I, someone with an inordinate fondness for bugs, have experienced fear when finding a wasp in my bed. But I also know how captivating, intriguing, and underappreciated all bugs can be. I aim to help others understand the positive aspects of bugs (even when those positive aspects are coupled with feelings of uncertainty). We should learn to at least appreciate these creatures, because without them, we would die. That isn't an exaggeration; it's just a fact. E.O. Wilson, an evolutionary biologist, said, "So important are insects and other land-dwelling arthropods, that if all were to disappear, humanity probably could not last more than a few months."[1]

For every single human, there are 1.4 billion insects.[2] For every one insect species that has been discovered, researchers estimate there are 10 unknown species. That means that although around a million species have been identified, 10 million unknown species may still be out there, waiting to be found. Invertebrates already account for about 95 percent of all species on earth. We as humans are seriously outnumbered. Thankfully, bugs are here to help. They support most of the food chain, they pollinate most of our food, and they decompose all that we don't want. Without them, we would quickly have no food and be buried in waste. They were here long before us (and dinosaurs), and they will be here long after we are gone.

Even knowing all this doesn't mean my faith in bugs hasn't been tested. On the day I found out I would have the privilege to write this book, I found a bed bug running across my pillow. For a while, I stood there stunned, having heard about the horrors of eradicating them, and not willing to believe that this was happening. As an apartment

---

1 Wilson E. O. (1992). The diversity of life. Belknap Press of Harvard University Press.
2 That doesn't even include other invertebrates like spiders, scorpions, and all ocean creatures without a backbone.

dweller, it was particularly worrisome: Did they come from another apartment? Has anyone else seen one and not known what it was? After capturing the bug, texting people who lived in the twelve other apartments in the building, contacting our landlord, and doing my own inspection, I sat down to breathe. What an amazing way to celebrate writing a book about bugs (that's sarcasm by the way)!

The bed bug experience continues to be psychologically exhausting: I still dry everything on high heat. I check the corners of the mattress multiple times a week. Our bed will forever be moved away from the walls. We have little bug catchers under the legs of the couch and bed. I constantly check every little piece of fuzz to be sure it is not one of them; however, this situation did lead me to learn more about them: their history, behavior, ecology. What I found was absolutely fascinating. Don't get me wrong, having bed bugs blows and I would never wish them on anyone, but the bugs themselves are evolutionary marvels (more on that in chapter 14).

My fascination with wildlife began when I was growing up between a large city and a decent swath of woods. I was always discovering something new, the good and the bad. I'll never forget pushing a dead raccoon into the creek and watching thousands of beetles abandon ship out of holes in the raccoon's skin. Growing up, I wanted to be an entomologist (bug scientist) and was even set on that throughout high school. In college, my career desires shifted. I never forgot about bugs, though, and finally went back to school for a degree in biology where I focused on insect conservation. With all my powers combined, I knew that I finally possessed all of the know-how and passion to write a book to let people know how and why bugs are just so cool (and why humans have thought so for centuries).

This book is for the curious bug hater or the undying bug lover. For those of you who dislike bugs, this book is to help you appreciate how impactful they have been to human culture and why they aren't such a bad thing. For those who already love bugs, the facts in this book will hopefully add to your already large amount of fascinating bug trivia. For everyone in between, this book will help you understand why we need bugs around, how they have defined culture, and how they will continue to be our saving grace in a world ravaged by human destruction. These little creatures need our help in order to survive because human interference has led to their decline. You don't need to love them to appreciate them or get a sense of their role in everything. (Of course, I'm hoping you love them a little bit after reading!)

The inspiration for this book tries to capture the ideas presented by Dr. Pritha Dey, a moth biologist. They talk about how animal numbers are dwindling and that using art/poetry may be the way to tell the insects' stories. For some species, those cultural works may survive longer than the animals will. Dr. Dey believes that together, science and art can be impactful:

> When effectively told, such narratives can provide the incentive to appreciate what is overlooked and inspire us to seek more knowledge—all crucial steps towards saving dwindling populations of insects before we lose them.[3]

Each section of this book is divided by types of bugs, typically by what are called orders (more on that below). Within each section are chapters about the different suborders or families within that order. So for the order Lepidoptera, there is a chapter on caterpillars, one on butterflies, and one on moths. Each chapter includes information on how each type of bug shapes our world, how we can interact with them, and how they show up in culture.

I've talked a lot about bugs, but what are they even? I think most people define *bug* like Supreme Court Justice Potter Stewart defined *porn*: "I know it when I see it." Similar to what is considered obscene, people have different ideas of what constitutes a bug. In this book, for the sake of brevity, I focus on insects (which are definitely "bugs"). However, it's important to note that other non-insect invertebrates are included in the world of bugs.

So how do I plan to define *bug*?

Most importantly, bugs do not have a backbone (called invertebrates). It's hard to tell whether or not they do just by looking at them, but mammals, reptiles, and birds all have backbones. Insects and other arthropods (another word for the backbone-less) can be identified by things like how many body segments, limbs, eyes, and mouthparts they have. The quickest way to determine what type of bug a spineless creature is may be by looking at the number of legs:

- 6 legs: it's an insect

3 Dey, P. (2022, Feb 21). Fluttering to the flame: Moths in art, literature, and poetry. Roundglass Sustain. roundglasssustain.com/wildvaults/moths.

- 8 legs: it's an arachnid
- More than 8 legs/no legs: probably a bug, but not an insect or arachnid

Even though we'll only be covering insects in the pages to come, I want to thank all the bugs that didn't fit in this book. Your time will come.

In the meantime, I want to dive into an important part of identifying bugs and reading this book: scientific classification, also called taxonomy. In elementary school, we made up our own mnemonic device to remember the different parts of this system:

Kangaroos **P**lay **C**ards **O**ut **F**ront, **G**o **S**ee

Kingdom **P**hylum **C**lass **O**rder **F**amily, **G**enus **S**pecies.

All life (that we know of) is placed into these categories created by western scientists, with kingdom as the broadest category and species as the most specific. Humans and insects are part of the kingdom Animalia, but we split off after that. Phylum Arthropoda contains all invertebrates on earth, which is where all of the bugs fall.

	Humans	**Bugs**
Kingdom	Animalia	Animalia
Phylum	Chordata	Arthropoda

Taxonomy is complicated and ever-changing as new discoveries are being made. The chapters may have information about different *suborders*. A suborder is a way to classify the creature within an order into smaller, more closely related groups, before you get down to the level of families. These levels aren't easy, and sometimes they shift. Arthropods, as you get more and more specific, are shifting frequently as more is learned about them and more species are discovered. Just know that *suborder* comes between *order* and *family*.

Here is a quick glossary of words I will use repeatedly that you may not hear every day:

- Arthropod: any animal without a backbone
- Insect: an arthropod with six legs and three parts
- Thorax: the middle of three parts on an insect
- Abdomen: the end (butt) of the three parts of an insect
- Tarsi: the equivalent of toes on bugs
- Mouthpart Types: chewing, siphoning (like butterflies), piercing/sucking (like mosquitoes), sponging (like flies)
- Exoskeleton: the hard outer shell of arthropods
- Exeuvia: the shed of an insect between some life stages, like the shell of a cicada
- Simple Metamorphosis: the bug looks the same throughout its life, just in different sizes
- Complete Metamorphosis: the young and adult look very different between when first born and into adulthood
- Larva: the first form for insects with complete metamorphosis
- Pupae: for insects with complete metamorphosis, the form in between the young and adult stage, kind of an adolescence, in which the insect does not usually feed or move around much
- Nymph: the young form of insects with simple metamorphosis
- Instar: a life stage of insects who do not go through full metamorphosis; often looks like smaller versions of their adult selves

NOTE: I would like to point out that while writing and researching this book, I saw articles almost daily that discussed some new super cool thing discovered about bugs. The facts in this book are only as current as early 2024. With a quick internet search, I'm sure you can find dozens of new things that they've discovered about insects in only the last month.

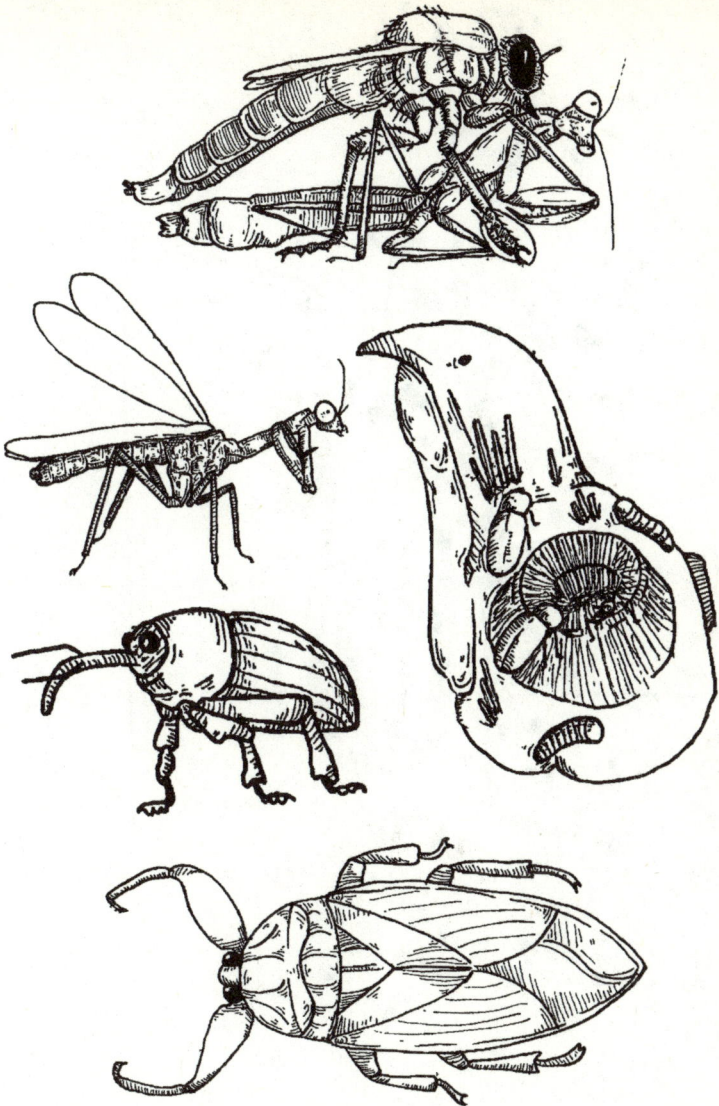

(top) robber fly (family Asilidae) eating a praying mantis; (middle right) turkey skull being cleaned by dermestid beetle larvae (family Dermestidae); (bottom) giant water bug (family Belostomatidae), also called the toe-biter; (middle left, top) mantidfly (family Mantispidae); (middle left, bottom) and acorn weevil (*Curculio glandium*)

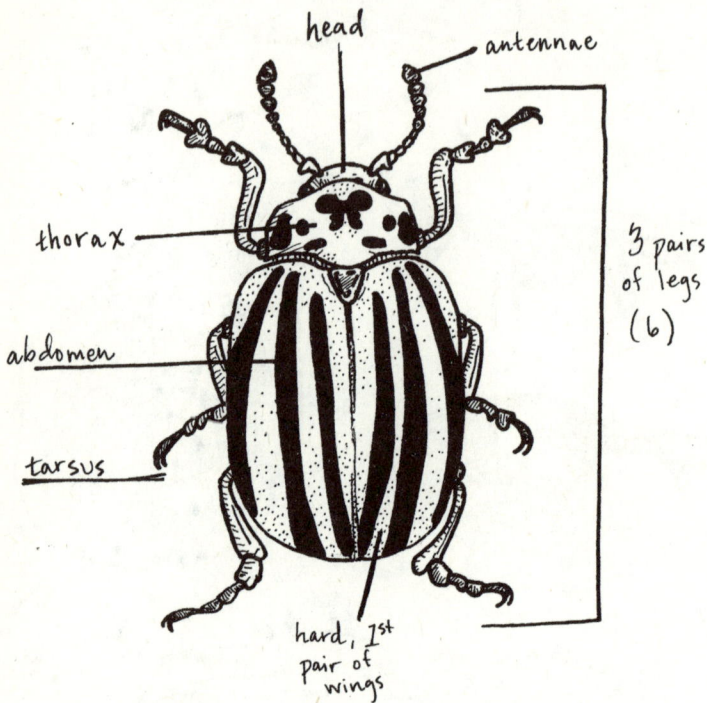

head

antennae

thorax

abdomen

tarsus

3 pairs
of legs
(6)

hard, 1st
pair of
wings

The insect shown here is what is commonly known as a Colorado potato beetle, *Leptinotarsa decemlineata*. As is common with insects, they have three pairs of legs, three body parts, two pairs of wings, and antennae.

The Evolution of Bugs

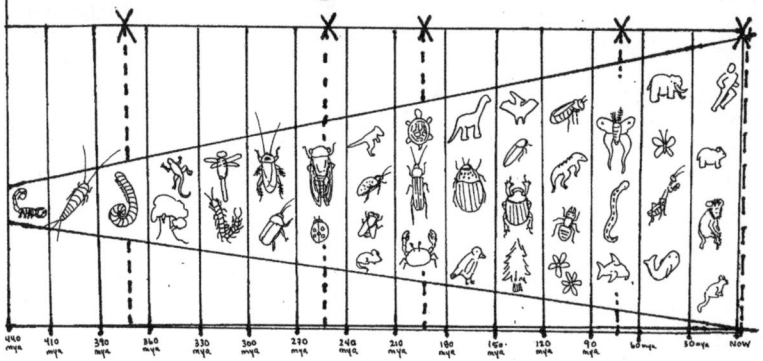

This timeline starts when the first arthropods were found in fossil records. Any "X" at the top of the timeline indicates a major extinction event, of which arthropods have survived five. We are currently in the fifth extinction event (the anthropocene).

According to fossil records, insects didn't start to show up until about 400 million years ago, and many species haven't changed their structure much since. This suggests that they figured out the perfect form and have stuck with it: Silverfish that you find in your bathroom look remarkably similar to their ancestors from hundreds of millions of years ago. Around 285 million years ago, bugs developed the ability to fly, which then created a large burst in their diversity. What I mean by that is, in combination with new types of plants showing up, numerous new insect species evolved and created more branches in their family tree. Butterflies and wasps didn't start to appear until around 200 million years ago. Mantises were first found in amber about 60 million years old. Insect bodies (exoskeletons) often break down before fossilization, so the fact that we have as many fossil records as we do is exciting. All this means that even some of the most recently evolved species were roaming the earth 59 million years before humans appeared. To put that into perspective, mammals had at that point been around for 300 million years and primates had been around for 20 million years. The first arthropods (trilobites), ancestors of insects, existed 515 million years before our most different hominid ancestors.

Bugs in Culture: Throughout Human History

With trillions of bugs running around, it is unsurprising that bugs show up repeatedly in human culture. The study of insects in culture, a field called "cultural entomology," is defined by entomologist Dr. Charles Houge as the "influence of insects on literature, languages, music, the arts, interpretive history, religion, and recreation."[4] In this book, I define culture in a broader sense. Yes, it includes language, music, art, and literature, but it also includes the norms and institutions of society, like medicine and pest management.

The earliest example of an insect recorded by a human is an image of a cricket that was inscribed on a bison bone around 20,000 years ago. Many origin stories across the globe include bugs as integral players in the creation of earth. In Cherokee lore, the world was originally covered in water and a water beetle dove under the water and brought mud to the surface, thus creating the first bit of habitable land.

In Cochiti stories, a darkling beetle held its abdomen in the air because it was given the responsibility of transporting a bag of stars. The beetle spilled the stars which created what we now know as the Milky Way. Embarrassment caused him to hide his head when approaching.

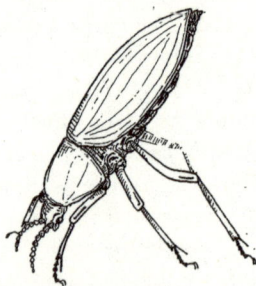

This is a common defensive stance for darkling beetles. They release a chemical to ward off potential predators.

As if the darkling beetle had not suffered enough, he was blinded and then had his eyes stolen from him. They were taken by a spider which explains why spiders have more eyes than insects.

In Mayan literature, yellowjackets were used as weapons by Mayan warriors and fireflies were used to guide two brothers who would become the sun and the full moon.

4 Hogue, C. L. (1987). Cultural entomology. Annual Review of Entomology, 32(1), 181-199.

Aztec, Greek, and Egyptian cultures have gods represented by or closely linked with bugs: the Aztec death goddess had moth wings, and Psyche, the Greek goddess of the soul, had butterfly wings; Khepri the Egyptian god of the rising sun was often depicted as a scarab beetle. People even named important places after insects: the Inca named their sacred valley "plain of the insect," Japan used to be called "dragonfly island," and the most important Aztec castle sat upon "hill of the grasshoppers." Christianity seemed to have different feelings about bugs because most of the representations in the bible are negative. Three of the ten plagues are bug-related, so that's a bummer: one of locusts to eat all of the food, one of lice or gnats to be an annoyance, and one of flies to bring disease to people as well as livestock.

All that said, bugs have continued to fascinate humans in a myriad of ways. They have been imbued with symbolism, inspired tech, and wormed their way into language. All of these topics and more will be discussed in greater detail in the pages of this book.

How Did We Get Here?

Unfortunately, in the twenty-first century we seem to be less knowledgeable and more disgusted by bugs than ever. Entomophobia is a condition that creates anxiety around insects. Part of it is most likely an evolutionary response: some bugs are dangerous to humans, so we have a natural fear. What people experience now may be an exaggerated form of that instinct, fueled by misinformation and lack of experience with bugs. Two Japanese scholars created an hypothesis called the urbanization/disgust hypothesis.[5] They determined that increased urbanization in conjunction with lack of knowledge and interaction has led to an increased disgust around bugs, mostly because the bugs we encounter in urban environments are pests. This leads to people equating any bug with a pest, even if the bug is actually harmless. People also have the idea of "their space" and a supposed separation from nature. Pests are considered pests because we said so, not because of any real definition. Pest control closely ties to our sense of power. We don't feel very happy when an animal adapts to "our space," even though we are the ones who created the situation for those animals. In the book *Pests*, the author Bethany Brookshire states that

5 Fukano, Y., & Soga, M. (2021). Why do so many modern people hate insects? The urbanization–disgust hypothesis. Science of the Total Environment, 777, 146229.

"pests are not irritating by nature. Instead, they are animal winners on a planet full of loss."

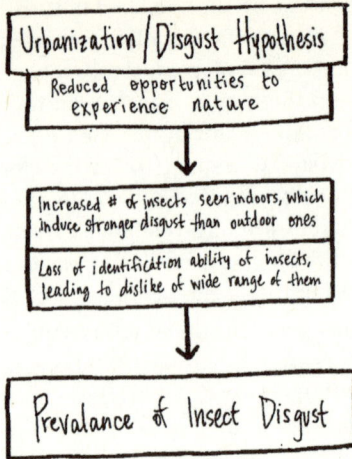

Urbanization / Disgust Hypothesis

Reduced opportunities to experience nature

↓

Increased # of insects seen indoors, which induce stronger disgust than outdoor ones

Loss of identification ability of insects, leading to dislike of wide range of them

↓

Prevalance of Insect Disgust

Adapted from the Fukano, Y., & Soga, M. (2021)

These common negative feelings have a genuine impact on conservation efforts around bugs. It creates a cycle of "We need bugs but don't like them, so we will try to kill them all, but we need bugs…" Conservation money is most readily given to animals that people feel connection or wonder with: giraffes, apes, elephants. The name for these types of species in conservation is "charismatic megafauna." These big animals (mega-fauna) share human traits or are mysterious/majestic, so we feel we have to protect them. Also, their disappearance is much more noticeable. Bugs may be charismatic but they don't often inspire connection. They have too many legs, alien-like features, and erratic movement. It is hard to see ourselves in them. Some entomologists use the term "charismatic microfauna" to describe the most charismatic of insects, like butterflies, dragonflies, and praying mantids, in an effort to establish connection across the chasm.[6][7]

People not liking insects isn't the only hurdle for funding bug conservation. Even for scientists, bugs are hard to find and track in the same way as larger animals. Also, insects don't live very long,

6 Bossart, J., & Carlton, C. (2002). Insect conservation in America: Status and perspectives. American Entomologist, 48 (2), 82-92. doi.org/10.1093/ae/48.2.82

7 Duffus, N. E., Christie, C. R., & Morimoto, J. (2021). Insect cultural services: How insects have changed our lives and how can we do better for them. Insects, 12(5), 377.

so tracking behaviors, movement, and overall numbers is extremely difficult. Organizations are less likely to fund a project if you can't tell them exactly what you're looking for. Not a great use of dollars in the opinion of the one who is funding. So, larger and seemingly more immediately impactful projects are funded.

The sheer number of bug species also makes the decision on what to study hard to narrow down. How do you focus on a single species when there are thousands of similar ones? Some organizations have done a great job of focusing on groups of bugs like pollinators. The movement around pollinator conservation has skyrocketed over the last ten years and has serious traction. Many of the factors that threaten pollinators threaten all insects: habitat loss, climate change, pesticide use, and disease. Addressing pollinator issues may help bugs overall. Nevertheless, more action is needed.

We currently know that over 1,000 species of bugs are considered vulnerable or critically endangered. There are 105 known extinct or possibly extinct species. It is possible that hundreds or even thousands of species went extinct before we even knew that they existed. Currently there are close to 2,000 species whose status is unknown. We know that these bugs exist but have so little research on them that we have no way to determine if they are endangered or thriving.[8]

Most bugs have ecological relationships that are essential for the survival of animals higher than them on the food chain. Salmon eat firefly larvae as their main food staple and grizzlies rely on thousands of miller moths to complete their diet. Co-evolution between bugs and plants over millions of years is also intimately connected. Wasps, bees, flies, butterflies, and moths evolved in conjunction with flowering plants 66 million years ago. Numerous plants evolved so closely with specific bugs that only those bugs can pollinate it. Some bugs evolved as parasites on individual species. Parasitoids, which are common amongst wasp species, are animals whose larvae are parasites on other animals, most frequently other invertebrates, that end up eventually killing their host. There are even parasitoids that parasitize parasitoids, up to three layers deep, like the movie *Inception* but with parasites. Without their particular host, they would die. So you can see, losing one species can lead to a series of breakdowns in the food web. Some

8 Dunn, R. R. (2005). Modern insect extinctions, the neglected majority. Conservation biology, 19(4), 1030-1036.

of those spots can be filled by other species, but when they are so specific, it could also easily lead to losing species forever.

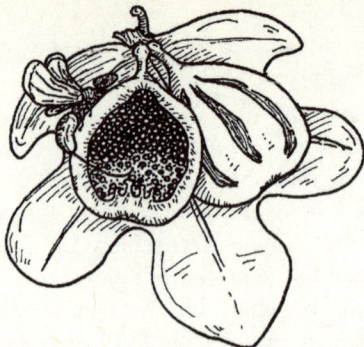

The female wasp lays her eggs inside the fig with her ovipositor.

As we get to the end of the introduction, I hope you are excited to continue delving deeply into bugs and their influence in our lives. Bounce around from your least favorite to favorite insects, or vice versa. Consume this content in any way you want and know that wherever your interest takes you, it will be deliciously fun.

Part I The Pretty Ones: Butterflies and Moths (Order Lepidoptera)

The cocoon of a sphinx, or hawk, moth. Family Sphingidae. May move or make a clicking noise if disturbed, as a defense against predators.

*I*n numerous studies, butterflies and moths rank highest on the lists of people's favorite bugs.[9] (Obviously not everyone loves butterflies, like my grandma, who has an intense fear of them. I'm unsure where the fear came from, but she tried to jump out of a moving car when one flew in the window.) They are a large part of the success of ecotourism—specifically ento-tourism: traveling to a place to see specific bug species. Because butterflies and moths are closely related, they share many similarities: both can be seen as agricultural pests; both are important pollinators and food sources; and both have flight patterns which could be described as chaotic or drunk. Some researchers believe they developed that flying pattern to confuse predators. (Watching birds try to catch a moth in midair can be wildly entertaining. You—and the birds—have no idea what move that moth is going to make next.)

Even with these similarities, moths and butterflies have a significant number of differences, mostly as they relate to physical appearance.

- Butterflies are active during the day; moths are active at night.

---

9 Kellert, S. R. (1993). Values and perceptions of invertebrates. *Conservation biology*, 7(4), 845-855.;

Shipley, N. J., & Bixler, R. D. (2017). Beautiful bugs, bothersome bugs, and FUN bugs: Examining human interactions with insects and other arthropods. Anthrozoös, 30(3), 357-372.; Skibins, J. C., Dunstan, E., & Pahlow, K. (2017). Exploring the influence of charismatic characteristics on flagship outcomes in zoo visitors. Human dimensions of wildlife, 22(2), 157-171.

- Butterflies have knobs at the end of their antennae; moths have large, fuzzy antennae without knobs at the end.
- Butterflies are usually brightly colored; moths are usually dull.
- Butterflies rest with their wings held vertically; moths rest with their wings flat.

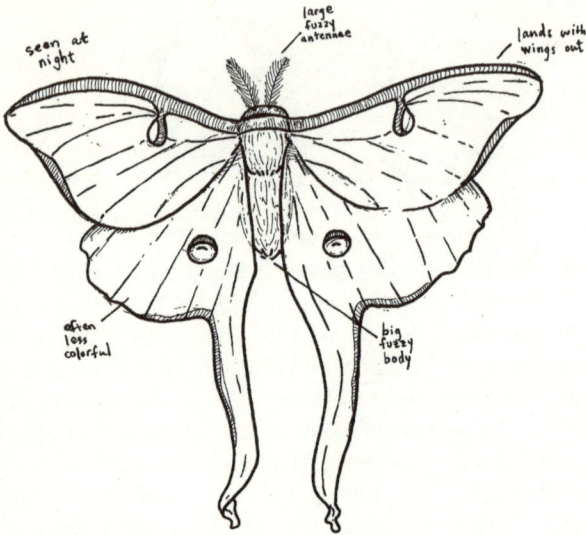

Luna moth (*Actias luna*)

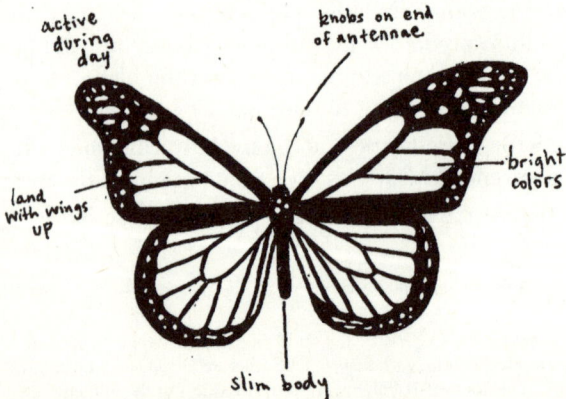

Monarch butterfly (*Danaus plexippus*)

While the above are often true, there are exceptions to the rule. The Madagascar sunset moth has all of the key aspects of a butterfly: brightly colored, thin antennae, thin body, active during the day. In fact, when first identified, it was categorized as a butterfly. Some years later, it was re-categorized as a moth. The giveaway? Its antennae are missing the distinct club that appears at the end of a butterfly's.

The order name for butterflies and moths is Lepidoptera, which means "scale-wing" in Latin. They were given that name because their wings, when observed up close, have scales. Most of the colors we see on an adult's wings, like the ones on the sunset moth, are not actually colored but are instead a reflection of light on the scales of the animal.

In this section, the chapters will discuss moths and butterflies in detail. Starting with caterpillars in chapter 1, since butterflies and moths both start that way, we'll discuss our fascination with them and why it is important to not touch them. Then, in the next two chapters, we'll dissect the cultural and ecological impact of moths and butterflies.

CHAPTER 1: CATERPILLARS (OF BOTH BUTTERFLIES AND MOTHS)

Before moths or butterflies can be adults, they have to go through their caterpillar phase.

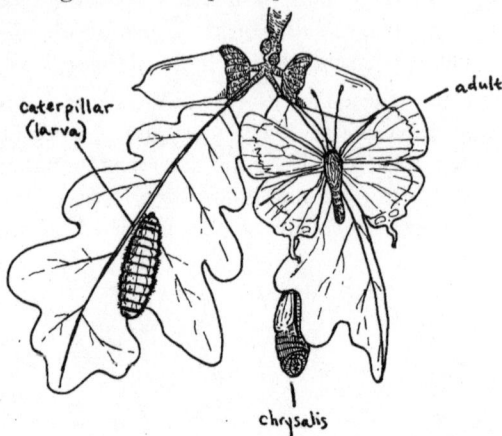

Life cycle of the Colorado hairstreak (*Hypaurotis crysalus*)

They begin their life chewing their way out of the egg. Some species completely consume the egg for sustenance. As caterpillars, their evolutionary MO is to eat enough food and survive long enough to make a cocoon (moth) or chrysalis (butterfly) and turn into an adult.

Before entering their cocoon/chrysalis, or pupal phase, caterpillars shed their skin and go through many instars, the life stages between molts. The time it takes for a caterpillar to turn into a butterfly/moth can vary greatly. Some species can make that growth in eighteen days while others may take upwards of ten months. These attributes are all tied to their food sources: if caterpillars have to wait long amounts of time for food, or fast, when temperatures dip, they can take much longer to pupate.

Along the journey to becoming adult insects, caterpillars have an anatomy that is full of wonders. Most breathe through spiracles located on the sides of their body. And while grown insects have six legs, caterpillars have "prolegs": little chubby sticky legs that help push the big bodied critters along. Evolutionary biologist Antonia Monteiro calls caterpillars "eating tubes" and states that they need "more legs to support their gut."[10]

Monarch caterpillar with prolegs. They will lose these in adulthood.

Cocoons and chrysalises, just like adult forms of lepidopterans, are often fairly easy to distinguish between. Cocoons (for most moths) are larger and fuzzy because of the silk used to weave them. The pupa sits inside the cocoon, but it is not attached inside. By contrast, chrysalises (for most butterflies) are smoother and almost an extension of the caterpillar. They are hard shells that form from the caterpillar's last molt and are their home for anywhere between ten days or ten months, depending on the species. Some versions of both cocoons and chrysalises are able to respond to external factors and wiggle or turn over as a defense. Some even produce noises to spook whatever is messing with the pupa.

How Caterpillars Shape Our World

Caterpillars are an extremely important part of the food web. Many animals, even humans in some cases, eat caterpillars. For example, these creatures make up the majority of the food that birds feed their young[11] and also serve as a food source for animals higher in the food chain. Besides serving as food, caterpillars also grow up to be important pollinators.

Still, some caterpillar species have a bad reputation since they are voracious eaters. Home gardeners may dislike the damage done by some species, especially hornworms on tomato plants. Thankfully, in

10 Pallardy, R. (2023, Oct 23). Caterpillars evolved their weird chubby little 'prolegs' from ancient crustaceans. Livescience.

11 "Working the Night Shift," National Wildlife Federation, nwf.org/Magazines/National-Wildlife/2018/April-May/Animals/Moths

places where the hornworm is native, there are wasps who parasitize them to help keep the number down. In places where these caterpillars are not native, humans have a harder time keeping them off of their plants. The good news? The damage hornworms do is not enough to completely kill the plant, so they get to eat and you get to keep your plant. And then some bird or wasp gets to eat them later!

You will see a pattern in each chapter: bugs are an essential food source for almost every omnivorous and carnivorous animal above them in the food web. Without them, food webs would completely collapse.

How Humans Can Interact with Caterpillars

Fuzzy caterpillars get a lot of love, which I totally understand. I remember playing with woolly bear caterpillars all the time as a kid. Other people share my fascination: there are woolly bear festivals in Ohio and North Carolina. In North Carolina, they even use this caterpillar to predict the weather for the upcoming winter. Tradition states that woolly bears have thirteen segments, which correspond to the thirteen weeks of winter—and that the amount of brown and black lines they have corresponds to the type of winter weather for each week: If the segment is black, there will be below average temperatures. If it is brown, temperatures will be average. There is no scientific evidence to back up this claim. Each woolly bear may have different markings in the same year, which probably correspond to camouflaging in their surroundings.[12]

But despite the understandable love toward fuzzy caterpillars like the woolly bear, it's important to know that picking up an unknown caterpillar can be extremely dangerous. A good rule of thumb? If you don't know what kind of caterpillar it is, DO NOT PICK IT UP. If it has spikes or warning colors, acknowledge the warning signs and admire it from a distance.

But some of the most dangerous may not have such obvious defenses. For example, he puss caterpillar looks like a weird, little moving toupee. I can completely understand the urge to touch it, but DON'T! The pain it can cause has been described as unbearable, with people writhing around on the floor for several hours in a space

12 "Groundhogs Are Old News: In This Tiny Town, Caterpillars Predict the Weather. Smithsonian Magazine, smithsonianmag.com/travel/groundhogs-are-old-news-in-this-tiny-town-caterpillars-predict-the-weather-180983146/.

somewhere between consciousness and death. No, you won't actually die from it (most likely), but they can cause seriously unpleasant reactions. Or take the giant silk moth in South America: it has been known to cause death, so seeking medical help is essential should you come into contact with a *Lonomia obliqua* caterpillar (to reiterate, they do not live in the U.S.).

These poisonous caterpillars are getting more and more air time, especially if they are introduced to a new country (like the puss caterpillar in the U.S.). The actor Jamie Dornan went on vacation in Portugal and was then hospitalized due to an encounter with a poisonous caterpillar. It was all over the news.

While our fear is understandable, poisonous caterpillars have always existed. Why the increased concern now? Because of human dwellings encroaching on habitats and climate change, places are seeing new species. Many of the species people are concerned about have always been around but perhaps less frequently seen. But none of these species are out to harm us. They have these defenses for a reason but are not pests simply because they exist. A healthy caution toward these creatures does not validate extensive fear and a push for eradication of certain species.

Apart from intense poisons, caterpillars have evolved other ways to avoid detection and survive into adulthood. Mimicry, looking like something else to trick the eye, is extremely impressive in caterpillars. Some look like twigs, some look like bird poop, and some even mimic snakes. You may see them in your garden and have no idea that they are caterpillars. In fact, regarding snake look alikes, numerous people post on the bug subreddits questioning whether or not they have a tiny snake eating their plants. The ones that look like snakes have little eye dots on the end of their body. Then, they hold up their body and move like a snake, swaying back and forth in a defensive position. Some of them even have a "tongue" (technically called an *osmeterium*) that pops out to further confuse a predator, as well as emitting a foul smell.

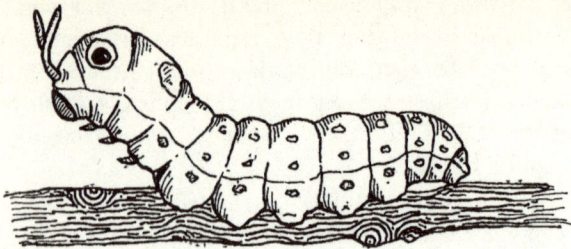

Caterpillar of the spicebush swallowtail butterfly (*Papilio troilus*) in full defense mode with "forked tongue" out.

They are very convincing to potential predators and even human passersby.[13] The ones that look like bird poop are crazy. They are giant swallowtail butterfly caterpillars. You'd never know they were a bug unless you saw them move. They mimic poop because it greatly reduces their chance to be eaten by their main predator, birds. They also live and munch on leaves so their placement seems natural. This allows them to avoid detection from all animals, including humans who may throw them off of their plants for fear of the plants being harmed. A phenomenal use of mimicry.

How Human Culture Is Shaped by Caterpillars

In popular culture, the idea of cocooning has made quite the impact. For example, many people use the word outside of the insect world to refer to rolling up in blankets where you can feel comfy and safe. In *Adventure Time*, the main character Finn sleeps in a cocoon-like sleeping bag. In the 1985 movie *Cocoon*, elderly characters reverse aging when they emerge from their cocoons. In Eric Carle's book *A Very Hungry Caterpillar*, his caterpillar goes into a cocoon and out comes a butterfly. (The majority of butterflies have chrysalises, but Carle stated that the word *cocoon* just fit better.)

Beyond cocoons, caterpillars have also influenced human culture through producing silk. Sericulture, the making of silk fabric, is a 5000 year old tradition. Silkworms (caterpillars of the *Bombyx mori* moth) create a cocoon of spun silk. Then they are boiled alive in their cocoons and the cocoon is unspun to collect the silk. Some people use wild varieties of moths, let them complete their life cycle, and then use the discarded cocoons. This approach is not often used since it makes

---

13 They also look a lot like the Pokemon character, Caterpie.

the process much longer. The majority of silk fabric comes from the death of tens of thousands of caterpillars. Do with that what you will.

Humans have also used caterpillars to wreak havoc in war. Insects were banned as a biological weapon in 1972. That ban only included bugs that were carriers of vectors given to them by scientists, like fleas purposefully carrying the plague. Untampered bugs were not explicitly banned. In the 1990s, the U.S. dropped caterpillars on coca plants in Peru in the hopes of disrupting the prolific drug trade. It didn't really work, but they tried, and those little caterpillars got to eat their hearts out. The problem (and sometimes the benefit) of using insects in warfare or other situations in which you want them to do a specific thing is that they can easily go to other plants or spaces to feed. They weren't just going to stay on the coca plants if other food sources were available. And they didn't.

Caterpillars love eating, and some humans love eating them. People from many countries have found that some of these non-poisonous species are delicious to eat. The emperor moth (also known as Mopane moth) has such yummy caterpillars that people in Africa are eating enough of the moth's population that its numbers are plummeting. Mexico has canned caterpillars in stores, and you may have heard the story of the "worm" in the bottle of tequila. Well, it's actually only in Mezcal (not all tequila), and that worm is actually the caterpillar of the giant skipper butterfly. In the past, distillers used caterpillars to determine how well preserved the alcohol was and if the concentration was satisfactory. Once you reach the bottom of the bottle, someone is supposed to eat the alcohol-soaked caterpillar. Their inclusion in bottles today is mostly for the novelty of it.

Then, there's the fascinating ways caterpillars show up in scientific inquiry. In research from 2014, Dr. Heidi Appel discovered that plants could identify when caterpillars are eating them and set up a defense in response. She played recordings of the sounds of caterpillars chewing on leaves, causing the plants to release mustard oil, a defensive chemical unappealing to insects. When she played the sound of an insect that did not eat that plant, the plant did not respond.[14]

---

14 Siegel, R. (2014, July 8). Plants Know The Rhythm Of The Caterpillar's Creep. *NPR*. npr.org/2014/07/08/329884061/plants-know-the-rhythm-of-the-caterpillars-creep

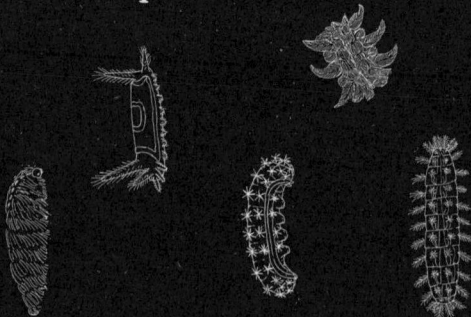

(in order from left to right) Southern flannel moth caterpillar, *Megalopyge opercularis*; Saddleback moth caterpillar, *Acharia stimulea*; Io moth caterpillar, *Automeris io*; Hag moth caterpillar (sometimes called "monkey slug"), *Phobetron pithecium*; Giant silk moth caterpillar, *Lonomia obliqua*)

If they make it through their initial stage of life by using their defenses, not being eaten by animals (which includes us), and munching on plants, caterpillars form a cocoon (moth) or a chrysalis (butterfly) to turn into their adult forms. The process is difficult to explain and comprehend. They basically liquify themselves into a lil' soup and then re-form as adult butterflies and moths. True transformation. No wonder their metamorphosis is used as an analogy for change and growth in a person's life. (I'm glad we don't liquify in adolescence!)

Bug Spotlight: Caterpillar Mimics

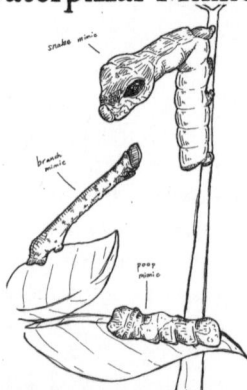

Caterpillar that looks like a snake (*Hemeroplanes triptolemus*, a type of hawkmoth); branch (*Biston betularia*, peppered moth); Caterpillar that looks like bird poop (*Papilio cresphontes*, Eastern giant swallowtail)

CHAPTER 2: MOTHS (MEMBERS OF 43 SUPERFAMILIES)

*M*oths showed up on the evolutionary timeline before butterflies, and their species continue to outnumber butterflies: 160,000 compared to 17,500.[15] Many moths coevolved with flowering plants and are continuing to do so. Mottephobia, moth phobia, is a real fear that is most likely caused by being "attacked" by a moth when someone was young. They do not attack people, but the surprise of them can be traumatizing for some. While some people are terrified of moths and some view them as pests, they are extremely important ecologically. While very few (around 1 percent) are considered pests, many species are important food sources or important pollinators.

How Moths Shape Our World

In the previous chapter, I mentioned that caterpillars were used to disrupt coca production. This is because they can be voracious eaters. And historically, the voracious appetites of moths have been involved in thwarting colonization. In Greenland, communities were abandoned because moths (probably armyworms also known as miller moths) were responsible for decimating their crops and pastures, basically starving them out. Since the Norse could not grow enough food, they abandoned their attempt at spreading to Greenland (though future colonizations by Denmark were successful). Moths tried to be fighters of colonization, so give them a little high five when you see them next.

Because of that extraordinary capacity to eat (along with other reasons), moths are often considered to be pests. But it's important to note that the most destructive ones are often introduced species (which means no natural predators), which is the fault of humans, not the moths themselves. Still, native species can also be viewed as pests. Miller moths stumble through Colorado in droves every year, causing many people to despise their presence. They are not destructive, just annoying. They are essential food for bats, bears, and other animals.

As another example of the impact of moths on the world, take the sponge moth. It was introduced to the U.S. in the late 1800s.

---

15 Smithsonian Institute (n.d.) Moths. si.edu/spotlight/buginfo/moths

These bugs are capable of eating large amounts of vegetation, so scientists released predators that would keep numbers of the moth in check; unfortunately, the flies and parasitic wasps they released (also introduced species) ended up targeting native moths. Now, some states ask that people participate in destroying sponge moth cocoons to help reduce their numbers.

For the few species that are considered pests (about 1 percent of moths), there are many that provide positive ecological benefits. Songbird numbers are directly tied to moth caterpillars. In years when moths are abundant, so are songbirds. When numbers of moths are low, numbers of songbirds plummet. Also, grizzly bears can eat upwards of 40,000 moths in a day. When they are preparing for hibernation and may not have access to vegetation for food, they eat nutrient packed moths which can have upwards of 65 percent body fat. Perfect for a bear that needs to bulk up.

While many moths are essential food sources for animals, moths can also be pollinators. Some use the "mess and soil" approach which basically means that they accidentally pollinate: they visit a flower, pollen happens to stick to their fuzzy bodies, and then they fly to another flower. Other moths are very effective and deliberate about pollination. Yucca moths are the only species that pollinates Joshua Trees, a type of yucca plant. Their caterpillars eat the fruit and seeds, and the adults feed on the blooms at night. Without them, yucca plants would cease to grow. Additionally, many plants have nectar that is only reachable by moths with long tongues, called proboscises. For example, hawkmoths have particularly long proboscises and are the only moth that can access the Prairie fringed orchid's nectar.

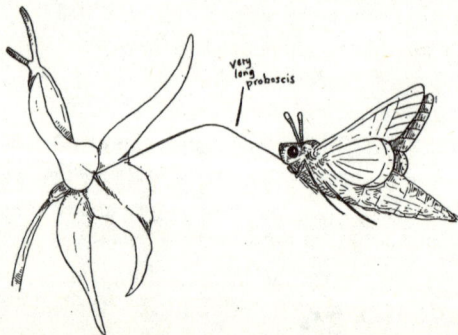

A hawkmoth feeding on orchid nectar that can only be reached by their long proboscis.

Moths are also important in the creation of vaccines. The armyworm moth is already used to create the yearly flu vaccine in the UK.[16] Researchers hope to use similar methods to continue producing covid vaccines. How does this work? Basically, there is an insect virus, called baculovirus. When the virus is present in eggs, the eggs themselves fight back by producing proteins to fend off the virus. Researchers have taken the antigens from covid or the flu and put those genes in the insect virus. The egg then creates a defense (called spike proteins) against covid/flu strains present in the egg. Researchers extract that defense to create vaccines for humans against covid/flu. In the UK, they have been using this method for the flu vaccine and are beginning to use it for covid vaccines. In the U.S., the fourth covid vaccine was created this way. I never would have guessed that moths could help with vaccines. What cool little dudes.

How Humans Can Interact with Moths

Have you ever encountered a moth in an unwelcome place in your home? Some species enjoy eating natural fibers, like wool, which means you can sometimes find them eating your clothes. They mostly like wool but will also eat linen, silk, or fur. Females will lay their eggs in those fabrics.

When I was growing up, we had a chest in the basement that had old blankets in it. Since we didn't open it much, my mom put mothballs in there to keep away bugs. The small balls release a gas that kills the moth and their larvae; however, the gas isn't great for humans.

Mothballs are used less these days because they have been found to have carcinogenic properties. We're willing to expose ourselves to deadly chemicals to avoid coming into contact with bugs. Easier and healthier solutions are available, such as washing and drying clothes/blankets in hot water, freezing the fabric, or putting up pheromone traps which only attract males, leading to lack of eggs.

How Human Culture Is Shaped by Moths

Symbolically, moths represent one side of a coin, with butterflies on the other. Moths are often associated with death and witches. Since moths are most often nocturnal, their association with darkness is understandable. They are the "dark side," which may also have

16 Nebraska Medical. (2022, July 21). Moths and tree bark: How the Novavax vaccine works. nebraskamed.com/COVID/moths-and-tree-bark-how-the-novavax-vaccine-works

something to do with their apparent suicidal tendencies. The phrase "like a moth to a flame" is commonly used to describe a person who is attracted to someone or something that may be dangerous for them. Its first recorded use in writing was in Shakespeare's *Merchant of Venice* in 1596: "Thus has the candle singed the moth." Some of you have probably witnessed moths dive-bombing your campfire or fire pit.

Of particular note here is the death's-head hawkmoth. While not native to North America, these creatures are very well-known because of their macabre-coded adornments and their inclusion in popular culture. These moths steal honey from nests, and people used to believe the myth that they had a deadly sting. Perhaps it is because they were often seen coming out of bee's nests. The skull-shaped markings on these moths did not help with superstitions either. I remember first seeing one on the VHS box for the movie *The Silence of the Lambs*. Edgar Allen Poe has a short story, "The Sphinx," in which a character has an encounter with a death's-head moth. In the story, the moth "emits a melancholy sound," which is false in real life. All in all, this moth's use in popular culture has been to represent death, darkness, and terror.

Outside of the macabre, moths have shown up in art, comics, and television. Mothra is a well-known combatant/friend of Godzilla in the films first appearing in Mothra vs. Godzilla (1964). She was created as a sympathetic character and her journey has continued through the films and into comic books. The villain Killer Moth appears in DC comics. Most moth-based comic book characters have very little in common with their real-world namesake, except the ability to fly. Understandable since moths don't have very formidable abilities. Moth-like Pokemon exist in the original 150, like Venomoth, and now even more moth-like characters have been created, like Frosmoth and Iron Moth.

Then, there is a famous cryptid of the U.S.: the Mothman. In 1966, a couple saw a large creature that was "shaped like a man, but bigger," but its most striking features were a pair of glowing red eyes and "big wings folded against its back." The legend has grown and continues to inspire people to travel to places where Mothman has been sighted.[17] The name Mothman was a reference to Batman and

---

17 Clarke, D. (2022). The Mothman of West Virginia. *North American Monsters: A Contemporary Legend Casebook*, 266.

has appeared in many TV shows and films. West Virginia now has a museum and a festival about this cryptid. Many drawings of the creature look like giant moths but some look like large birds. Some people believe it could have been a sandhill crane that was spotted by the original couple, as sandhill cranes are very large birds with eyes that "glow red" when light is shown on them. The glowing red eyes are not found on moths, but the idea persists.

Personal interpretation of mothman with bird legs

Quick Facts: Moths

- They are often easy to distinguish from butterflies by a larger, fuzzier body and wings that lay flat when they land.
- Some species don't have mouths as adults, like the luna moth, and rely on the food they ate as caterpillars.
- They are useful in the creation of vaccines for illnesses like COVID and the flu.
- They are essential for the creation of silk but are killed in the process.
- Most moths are important as a food source (like the miller moth) and as pollinators (like the yucca moth).

Bug Spotlight: Death's-head Moth, Genus *Acherontia*

Death's-head moth in caterpillar form and adult form, as well as an adult eating honey from a beehive

- They don't sing or make any noise.
- They eat honey out of beehives and mimic the smell of bees so that they aren't discovered.
- If they are discovered, they have a thick waxy substance that helps them not get stung.
- They appear in popular culture as omens of death, but they are harmless to humans.
- There are three known death's-head moth species, all of which are nocturnal (which means active at night).

Chapter 3: Butterflies (Superfamily Papilionoidea)

One of the most well-liked groups of insects is butterflies. They make people think of springtime. They are associated with life, growth, and transformation. Unlike moths, butterflies are most likely to be seen by humans during the day. Similar to some moths, they are pollinators using the "mess and spoil" method and are often found on colorful flowers. They themselves are known for their bright colors. They split from moths on the evolutionary tree about 60 million years ago when they decided to go for plants that were only available during the day. Since they are loved by a majority of people, they have been a great bug mascot for conservation efforts—a perfect example of a charismatic microfauna. Because when butterflies start to disappear, people notice.

How Butterflies Shape Our World

Positive portrayals of butterflies in popular culture match their positive ecological and economic impact. As previously mentioned, they are associated with life and their symbolism can run very deep in some cultures. In Mexico, monarch butterflies are seen as the spirits of deceased loved ones and are thus treated with respect.[18]

But the monarch's impact doesn't end there. The event that is the western monarch migration[19] begins in southern Mexico and makes its way through Arizona, Nevada, Oregon, Washington, Idaho, Utah and California. The migration starts in March and finishes in Mexico around the end of October. Then, the monarchs overwinter for about four or five months before heading back out on their migration in the spring. Many generations (around four) are born and die throughout the journey. These migrating monarchs live longer than those who don't migrate: six weeks versus two weeks.

---

18 Merroto, T. (2022, Oct 31). *Winged messengers: How monarch butterflies connect culture and conservation in Mexico.* Smithsonian Institute. folklife.si.edu/magazine/monarch-butterflies-mexico-culture-conservation

19 Quick note: Not all monarchs (*Danaus plexippus*) migrate such great distances. There is one type in Florida that doesn't migrate at all. The monarchs' range is actually worldwide with few species in other countries not migrating great distances.

Mexico, particularly the state of Michoacán, relies on monarchs for tourism as thousands of people come to see the migration every year. Without butterflies, towns whose businesses rely on ento-tourism would suffer. Almost every U.S. state has a monarch festival every year in anticipation of the monarchs on their path. This is especially true on the west coast.

Mexico has been working on monarch conservation efforts for years. In the U.S., organizations have conducted counts of monarchs throughout the season. Since the 1980s, the number of monarchs went from millions to just 2,000 in 2020.

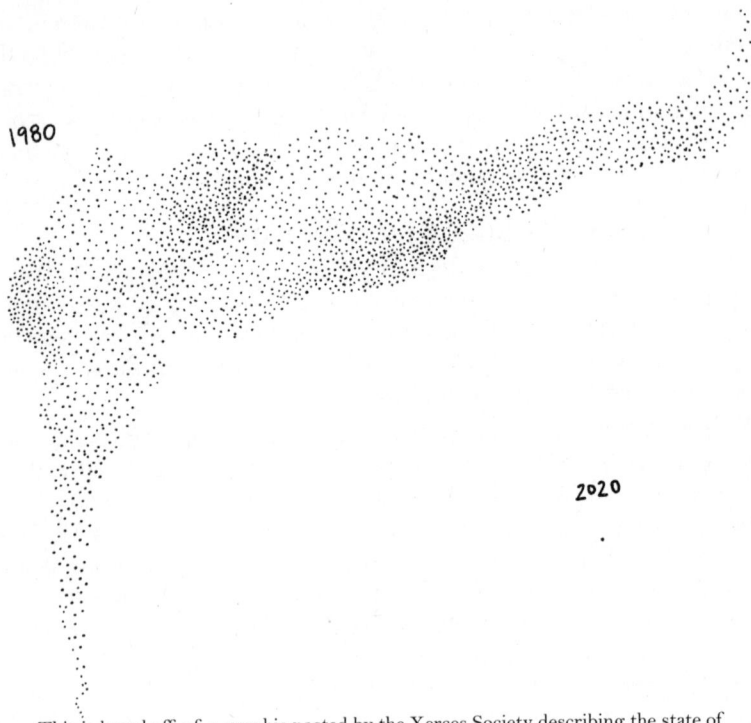

1980

2020

This is based off of a graphic posted by the Xerces Society describing the state of Monarch numbers recorded in 2020. (Original graphic appeared on their Instagram page on January 19, 2021.)

2022 saw an increase in monarch numbers, but numbers were down 30 percent in 2023. The National Wildlife Federation created the Mayor's Monarch Pledge to get cities on board with pollinator conservation across Mexico, the U.S, and Canada. 365 cities signed the pledge in 2023. Even with all of this conservation work, numbers are 90 percent lower than they were in the 1980s. Scientists have identified a few reasons why these declines are happening and how we can fight for the monarchs (more on that in the next section).

How Humans Can Interact with Butterflies

In the 1950s, a blue butterfly became famous for all the wrong reasons. The Xerces blue butterfly's home, the coastal sand dunes of the San Francisco peninsula in California, was taken over by development. Since the species lived in such a specific location and relied on that area, they went extinct. The caterpillars of the butterfly exclusively ate plants that belonged to two different genera that were only found in that area. Some insects are generalists while some rely on a very few types of plants. Those insects who rely on specific plants, like the Xerces blue, are at higher risk of extinction should those plants disappear, and the insect isn't able to adapt to eating other plants.

The first sighting of the Xerces blue was in 1852 and the last sighting was in 1943. It is the first species in the U.S. to have gone extinct because of urban development. As a response to this event, The Xerces Society was created. The organization is an invertebrate conservation society that named itself after that butterfly. They do educational outreach in the name of the extinct butterfly in the hopes that we can avoid that fate in other insect species.

Researchers believe the decline in butterflies, particularly monarchs, is the result of several issues: lack of food source (since they mostly eat milkweed), habitat destruction, and climate change. Similar to the Xerces blue, without food and habitat, numbers dwindle. Pesticides used on common crops, like corn and soy, kill most other plants around them, including milkweed. Some seeds are treated with neonicotinoids[20] which are also dangerous to butterflies and other pollinators. Issues related to climate change, like harsher winters and erratic weather, lead to higher numbers of die-off while overwintering and unevenly timed blooming of flowers/emergence of larvae, leading to butterfly starvation. Sites where monarchs overwinter, which they need to keep them safe and secure until spring, have been dramatically reduced due to logging and urban development in Mexico and California.

20 Xerces Society. (n.d.). Understanding Neonicotinoids. xerces.org/pesticides/understanding-neonicotinoids

Without food and shelter, these butterflies cannot survive. Thankfully, in 2022, monarchs were presented as an insect that should be placed on the endangered species list. Numbers are slowly increasing due to current conservation efforts. There is still a lot of work to be done but we are moving in the right direction.

If you are able, please plant native milkweed plants (and other native pollinator plants) in your area for their use. Don't know which milkweed is native to your state or region? Check out this table:[21]

Most Common Native Milkweeds and Which States to Plant Them			
Showy Milkweed (*Asclepias speciosa*)	Butterfly Milkweed (*Asclepias tuberosa*)	Swamp Milkweed (*Asclepias incarnata ssp. incarnata*)	Green Antelopehorn Milkweed (*Asclepias viridis*)
Arizona	Arkansas	Colorado	Arkansas
California	Florida	Idaho	Kansas
Colorado	Iowa	Iowa	Louisiana
Idaho	Kansas	Kansas	Mississippi
Montana	Louisiana	Minnesota	Missouri
Nebraska	Minnesota	Missouri	Oklahoma
New Mexico	Mississippi	Montana	Texas
Nevada	Missouri	Nebraska	
North Dakota	Northeast	North Dakota	
Oregon	Mid Atlantic	Oregon	
South Dakota	Southeast	South Dakota	
Utah		Washington	
Washington		Wyoming	
Wyoming		Southeast	
		Northeast	
		Mid Atlantic	

Monarchs are unique in that they show the cultural, economic, and ecological importance of insects in a single species. They are not the only species of butterfly that are important for those reasons. Other types have been important in positive ecological and economic changes in places like Papua New Guinea. Captive breeding programs have replaced environmentally destructive practices like logging and created economic value in existing native fauna. Packaged insect

21 Information obtained from Xerces' Milkweed Guides (xerces.org/milkweed/milkweed-guides). For more information on where to buy seeds and why to buy bee-friendly varieties, visit xerces.org/pollinator-conservation/native-plant-nursery-and-seed-directory.

specimens for pinning in the U.S. are extremely popular and can be purchased from many websites and stores. Many shops claim that their specimens die of natural causes and are sustainably sourced; however, there is little oversight in sourcing accuracy. In 1978, the Insect Farming and Trading Agency was created by the Papua New Guinea government to address exploitation of their native fauna. This organization is now defunct. A study from 2023 found that just on eBay, over 50,000 butterflies were sold in a single year from over five hundred sellers in over forty countries.[22] Some of these species were listed as threatened or endangered. The International Union of Conservation of Nature had much lower numbers of trade reported.

As we see, business is booming, and people profit from the moths and butterflies that are pinned by hobbyists, artists, and researchers worldwide. Since there is no specific organization overseeing the validity of these claims, the legitimacy of their sustainability will need to be taken in good faith. Particular places, like Butterfly Pavilion (an AZA accredited invertebrate zoo in Denver), have created their own farms with local connections in countries like Madagascar where they source many of their butterflies and moths; however, the oversight in other organizations is less transparent.

Some people and organizations have taken the dwindling numbers of butterflies as a call to conservation action through rearing and releasing butterflies back into their native habitats. Scientists warn against large-scale captive breeding releases for a few reasons. They believe the potential impacts could be negative, such as transmitted diseases between captive and wild caterpillars, overall less fit, and loss of genetic diversity (if closely related individuals inbreed).[23] People continue to release captive butterflies against the words of those scientists. In 2024, a California wildlife organization is releasing endangered caterpillars to a region that has seen a rapid decline in the butterfly numbers. The effort is coming from a good place, but some scientists may call the effort misguided. Humans will try to help in situations where they think they know what they're doing but might not. I used to think that having honeybee hives was a conservation effort (more info on why that is not

22 Wang, Z., Chan, W. P., Pham, N. T., Zeng, J., Pierce, N. E., Lohman, D. J., & Meng, W. (2023). One in five butterfly species sold online across borders. Biological Conservation, 283, 110092.
23 Altizer, S. et al. (n.d.) Joint statement regarding captive breeding and releasing of monarchs. Xerces Society. xerces.org/monarchs/joint-statement-regarding-captive-breeding-and-releasing-monarchs.

the case in chapter 9). Thankfully, we can learn from our mistakes and new research, then adapt our efforts.

How Human Culture Is Shaped by Butterflies

Throughout history, butterflies have found themselves in art, jewelry, and literature. One of my favorite short stories ever, "A Sound of Thunder" by Ray Bradbury, shows how important something as small as an insect can be. If you haven't read it, go find it now. It's great. A butterfly changes the course of history. The concept is a take on chaos theory or butterfly effect. This theory was made famous to viewers of *Jurassic Park* when Malcolm (Jeff Goldblum's character) says "A butterfly can flap its wings in Peking, and in Central Park, you get rain instead of sunshine." Or take the very popular children's book *The Very Hungry Caterpillar*, which shows a caterpillar turning into a beautiful butterfly. Then there's Psylocke from the X-Men team, who has projections that are shaped like butterflies (mirroring Psyche, the Greek goddess associated with these magnificent creatures).

Linguistically, everyone has had "butterflies in their stomach.". That expression first appeared in writing in 1908. Author Florence Converse wrote "The 3 o'clock train gave him a sad feeling as if he had a butterfly in his stomach." We also have the butterfly stroke in swimming, and someone can be a "social butterfly".

Colorful butterfly wings have been used by humans as inspiration for art and jewelry for centuries. Many of the colors that people notice, like the bright blue of the blue morpho butterfly, are not actually the color blue. The trick of the eye is called "structural coloration." Their wings are made with tiny scales and those scales have tiny structures that reflect certain colors. (Hence, the scientific name of lepidoptera for the order: *lepidos* means "scale" and *ptera* means "wing.") The blue morpho's structures reflect a blue light, whereas the structures of other butterflies may reflect an iridescent green. Whatever the reflective color, the butterfly does not contain the colored pigment. Blue is rarely found as a pigment in nature. The obrina olivewing, a butterfly found in South America, contains actual blue pigment, which is extremely rare. Many animals get their pigment from the food they eat, and blue is rare in plants as well. Humans love the color blue, but it isn't an easy color to make, even in the insect world.

Quick Facts: Butterflies

- They are often thinner than moths and land with their wings up.
- They are associated with beauty and life (the opposite of moths).
- Many butterfly species are having issues with dwindling numbers, like the monarch butterfly.
- If you want to help the monarchs and are on their migration route, plant some milkweed.
- If you want to help other butterflies, plant butterfly-friendly plants and leave the stocks after the end of the year.

Bug Spotlight: Xerces Blue, *Glaucopsyche xerces*

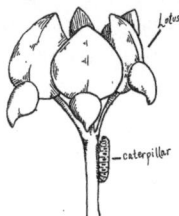

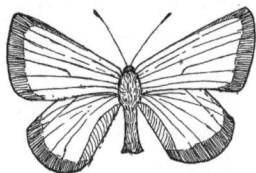

Xerces Blue: adult form and caterpillar form on a lotus plant (which was its main food source)

- They lived in a very specific part of California, in the sand dunes.
- Humans wanted to develop that area and cleared the area.
- The plant that these bugs fed on, a species of lotus flower, no longer grew in the area.
- Because of this, they went extinct.
- They are believed to be the first butterfly species to go extinct because of human interference.

Part II The Ones That Show up Out of Nowhere: Flies (Order Diptera)

I remember watching my mom kill a fly on the counter and as it died, it laid eggs on our counter. It was crazy watching something living come out of something dead. And during breakfast no less.

Flies are one of the few animals that appear on every continent. They are successful in all human environments, including science outposts in Antarctica. I'm talking specifically about the house fly, whereas some flies have more specific parts of the world in which they are found. (There is a house fly buzzing around my desk right now and I have no idea where it came from. It keeps landing on my arms and annoying me. Now it's just hanging out on the ceiling, upside down with its sticky little feet.) In short, they are the most ecologically varied insects simply because they live everywhere humans do. Some entomologists believe that the common house fly would not survive without people. Our relationship with them is called *commensal*: they benefit from us but we don't really benefit from them, nor are we harmed by them. That isn't true for all insects in their order, Diptera, since there are quite a few we benefit from, as well as a few who are harmful to humans.

Diptera means "two wings." Looking closely at a house fly, you'll notice that they only have two (whereas many insects have four). If you look very closely, you'll see two little nubs next to their wings. Those are the remnants of non-flight wings and are called *halteres*. Much about halteres is still a mystery, but it has been documented that the halteres are essential to flies' abilities to fly and maneuver as they do.

In addition to their flight skills, flies have a 360-degree view of the world and taste with their feet. In both the adult and larval stages, they are extremely important decomposers. As it turns out, many of the least-loved insects are decomposers. Without them, we would truly be covered in all the decaying things. They are also the second most important order of pollinators after bees (order Hymenoptera).

Some species of flies evolved with plants as important pollinators and created some highly specialized relationships. The chocolate midge is the only pollinator for cacao plants.

actual size
1mm

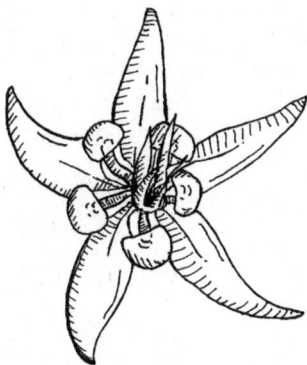

A chocolate midge, whose actual size is 1 mm, and a cocoa flower. The opening in the flower is so small, the midges need to be small enough to fit. They are the only ones who can.

Without them, we would not be able to have cacao plants and therefore, no chocolate. I definitely never thought I'd have to thank flies for my dark chocolate bar.

Most flies are harmless and extremely important, some are pestering (but also important in their own way), and very few cause serious health issues. Some can be beneficial and also annoying, or deadly but beneficial. Others are just harmless. They can fall into one or all three categories.

NOTE: The chapters in this section do not have suborders or families for the types of flies being discussed. They are grouped by factors outside of taxonomy.

CHAPTER 4: THE HARMLESS AND OFTEN UNNOTICED FLIES

*T*he majority of flies fall into the harmless and/or misunderstood category. The flies in this category may be predatory of pests, important in pollination, not seen in houses or swarming (therefore not pests), or just overall truly harmless. Physically, these flies can look drastically different—from large and delicate to tiny and fuzzy. Some even look pretty badass. Predatory flies, like robber flies, are rarely seen. When they are, people truly don't know what they are looking at. Is it a wasp? Or like a mutant fly? They are large and pretty badass looking.

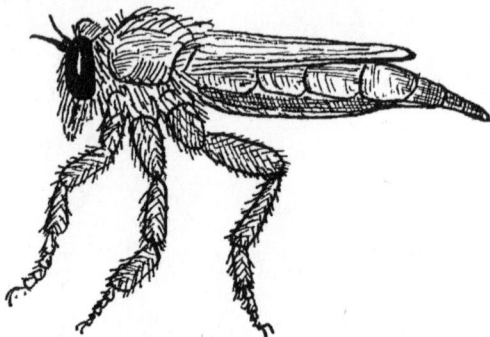

Robber flies, sometimes called assassin flies, can be up to two inches in length and feed exclusively on insects. They appear on every continent except Antarctica.

They have a fairly different stance to the house fly. Robber flies have fairly long legs with a bristly, svelte body. Growing to upwards of two inches, they are skilled hunters who catch their prey mid-flight. Their prey is other bugs who fly like flies (including ones we consider pests): bees, wasps, or beetles, you name it. Humans are most definitely not on their menu.

Some robber flies mimic other insects to appear unappetizing to predators, like birds. There is a robber fly who has the markings and smooth appearance of a paper wasp. Unlike some of their fly brethren, they have a piercing-sucking mouthpart. They do not bite or sting humans, however.

Robber flies look tough because they are. Their size helps to get that idea across. There is another fly that is a similar size, or even bigger, that is often misunderstood. The crane fly, while the same size as a robber fly, has a very different body type. They are a large, delicate fly that looks like a giant mosquito. Spoiler alert: It isn't. Check out the spotlight at the end of this chapter for more info. It is completely understandable to think that it may be a mutated mosquito. While they are related to mosquitos, they are not remotely similar, so never worry about them biting or stinging you.

How Flies Shape Our World

While we think of flies hovering around poop and decay, only their larvae like to eat those things. And not all fly larvae eat decay. Some eat other insects, particularly aphids. Karl Wollton, an ecologist in the UK, believes that the hoverfly larvae in the UK eat 6,000 tons of aphids, or 20 percent of the population, every year. This makes them important for pest control, since aphids are known for their crop destruction.[24]

As mentioned above, flies are second only to bees as the most important pollinators. The two largest groups of pollinator flies are hoverflies and blowflies. Adults are often attracted to nectar and pollen (with a preference for the putrid-smelling). Both are generalists, meaning they like all types of species and don't rely on a specific flower. They eat nectar for energy and pollen for growth. In a study that collected data about pollinator visitors to the 105 most planted crops in Australia, New Zealand, and the U.S., researchers found that "hoverflies visited 52 percent of the crops and blowflies visited around 30 percent."[25] Those numbers are huge!

In fact, hoverflies (also called flower flies or syrphids—from the genus *Syrphidae*) are second only to bees in pollination. With their black and white stripes or other distinctive markings, many of them ward off predators by mimicking bees or wasps. If you are trying to determine the difference, look for two wings and a distinct head shape/eye location.

24 Pain, S. (2021, Mar 8). How much do flies help with pollination? Smithsonian Magazine. smithsonianmag.com/science-nature/how-much-do-flies-help-pollination-180977177/
25 Rader, R., Cunningham, S. A., Howlett, B. G., & Inouye, D, W (2020). Non-bee insects as visitors and pollinators of crops: Biology, ecology, and management. *Annual review of entomology*, 65, 391–407.

Hoverfly
Syrphus ribesii

Honeybee
Apis mellifera

Yellowjacket
Vespula pensylvanica

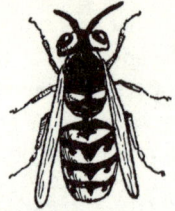

The similarities between this fly, bee, and wasp help to ward off predators even though the fly cannot sting. You can tell the difference based on the head shape, as flies have much larger eyes that extend from the front to the back of their head.

Blowflies are also important pollinators, which may be surprising. They are often talked about for their negative effects; however, they are effective at transporting pollen from flower to flower. Their big, fuzzy bodies can carry a lot around. However, unlike bees, they don't have to collect pollen for their young or nest so they can fly wherever they want. Some travel very long distances and can do so at varying times. They can do so because they aren't beholden to their young, which do just fine eating some dead stuff or aphids.

Another asset toward pollination? Flies have higher thresholds for cold weather and adverse weather conditions like rain or wind. Adding to the list of pollination advantages, they can breed quicker than bees. So, if you are looking to attract flies to your garden, scientists suggest planting white flowers with easy pollen access, like cow parsnip or frost aster (if it's native to your area).[26] Strange smelling flowers are also a good bet, like the Eastern skunk cabbage (native to eastern U.S. and needs lots of water) or the corpse flower.

How Humans Can Interact with Flies

Scientists are trying to determine if farmers can harness these flies to do their pollinator bidding. From recent studies, researchers have found that flies are much more tolerant of enclosed spaces like greenhouses, compared to bees. Apparently, bees get pissy in those spaces and are more likely to sting. Flies don't sting and are often found

26 Tooker, J. F., Hauser, M., & Hanks, L. M. (2006). Floral host plants of Syrphidae and Tachinidae (Diptera) of central Illinois. *Annals of the Entomological Society of America*, 99(1), 96–112.

in clusters anyway. The difficult part about using flies as purposeful crop pollinators is that farmers would also need to keep carrion around or a supply of aphids or a pond that has decaying things in it. Without a place to lay their eggs, adult flies won't spend much time in a grove or field. And researchers who are currently conducting fly studies say that the smell is unbearable. Mango blossoms already smell strange enough, then add in the smell of dead animals. Not necessarily ideal.

One fly-plant relationship is highly specialized and very connected to us as humans. What would the world be like without chocolate? That's a heavy question for many reasons. I also know a surprising number of people who do not like chocolate (I'm looking at you Leslie E.). Others make up for that and then some by spending upwards of 25 billion dollars, in the U.S. alone, on chocolate. None of that would be possible without a very special fly: the chocolate midge. A few species in the (unpronounceable) subfamily Forcipomyiinae are responsible for cocoa pollination. The midges themselves are extremely small, sometimes called "no-see-ums". Their tiny size allows them to access the small cacao flowers.

Like many insects, the future of the midges is of great concern. Chocolate is a profitable business, but the process is a delicate balance: flowers are only open a few hours a day, and then only around for a day or two. The traditional setting for growing cacao, undergrowth trees covered by thick canopy trees that create a humid environment, is in danger. Plantations remove the canopy to plant more trees which creates a drier environment, which is less desirable to chocolate midges. Our want for chocolate outpaces what can be produced (not to mention the inevitable slave labor and child slave labor that occurs on many farms). It's all bad. At the current rate, unless practices change in a myriad of ways, we may be without chocolate at some point in the future. Only when humans address climate change can they make steps towards saving the fly and the chocolate.

How Human Culture Is Shaped by Flies

In environments that can be more easily controlled, the larval form (maggots) and adult forms of another harmless fly have been important to scientific discoveries, specifically genetics. The fruit fly (*Drosophila melanogaster*) has been used for over a century in research. It made landfall in the United States when the fruit trade began expanding in the 1800s. I'm sure many of you had to breed fruit flies in high school or college biology. They reproduce very quickly. Because of this, you can see genetic variations much faster since they go through up to 25 generations in a year. They also share 60 percent of our DNA, but more importantly, 75 percent of the genes we share cause diseases in humans, meaning that they can help us address diseases in ourselves. 25 generations a year is a lot more to work with when compared with human generations. Researchers completely decoded the fruit fly genome in 2002, leading to additional breakthroughs in genetics.[27]

Starting in 1910, fruit flies were mostly used to determine inherited traits from parents. By the time we got to the 1950s, scientists began using them in the lab to evaluate physical development and pattern formation. This led to the understanding of Hox genes, which are the genes that specify what parts go where in an embryo. For insects, the Hox genes determine which appendages go where. It's like a toolbox. That translates to humans as well: where our arms, toes, etc. are gonna grow. Scientists found those genes in flies and were able to then identify them in other animals, including humans.

So culturally relevant is this discussion when it comes to flies that there is an X-Files episode (titled "Post-Modern Prometheus") about a geneticist who researches fruit flies and uses that knowledge to create a genetically mutated human sibling. Even if it does exaggerate the plausibility of creating genetic mutations in humans in the same way as in flies, it's a great episode.[28] Sometimes, house flies are used in research because they are much easier to physically handle than tiny, tiny fruit flies; however, fruit flies are still the gold standard for laboratory research.

They also helped with discoveries in behavioral neurogenetics, with studies in circadian rhythms, learning, and memory. Addressing "specifically human" issues like alcoholism and other addictions has been possible with fly brains and has helped determine which parts of the brain are affected by continued usage of certain substances. Currently, scientists are using fruit flies for research on Parkinson's, aggression studies, and sleep studies. Vertebrate animals owe a lot

27 Bellen, H. J., Tong, C., & Tsuda, H. (2010). 100 years of Drosophila research and its impact on vertebrate neuroscience: a history lesson for the future. *Nature Reviews Neuroscience, 11*(7), 514–522.

28 Genetic mutation doesn't really work like that, but genetic discoveries can make people nervous, in relation to designer genetics. Watch *Gattaca* for an idea of what that means.

to fruit flies. They will continue to be the gold standard in genetic research for decades to come.

Bug Spotlight: Crane Fly

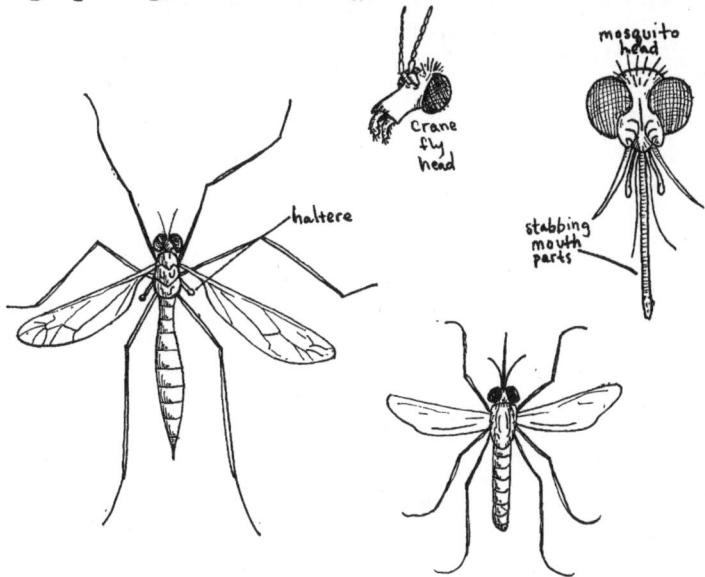

Crane fly (left), can be as large as 1.5 inches; their halteres (where a second pair of wings once were) can be seen if you look very close. Mosquito (right) are much smaller at an average of .4 inches. Notice on the crane fly head that they don't have stabbing mouthparts like mosquitos, therefore cannot bite humans.

Crane fly adults don't even have mouthparts. Their abdomens may wiggle around, giving the impression of trying to sting, but they can't sting. Female crane flies have an ovipositor they use to lay eggs. It may look stinger-esque, but it is only a small tube used to deposit eggs.

Some people call them "mosquito hunters" or "mosquito killers," but the adults can't eat at all, and the larvae are herbivores. The mosquito killing falls to the robber fly (and the birds and the bats).

Crane flies are also very sensitive to pollution, which can lead to them being good indicators of water quality. The larval stage does not tolerate pollution well and adults will not lay eggs in areas with high levels of pollution. When I say high levels, to us, the "level" of pollution may not be noticeable. Their sensitivity is much higher than ours.

Thankfully, most flies fall into the harmless category, but we are less likely to notice them since we do not interact with them as often. Of the over 150,000 species of fly, 99 percent are harmless and beneficial. Unfortunately, the ones we notice the most are annoying or potentially deadly, even though the total number of species in that category is about ten.

Chapter 5: The Annoying but Not Quite Deadly Flies

When I was a kid, my family would spend a lot of time camping and going on float trips. At dusk, when we'd be swimming or hanging out at the water's edge, the damn horseflies would come out in force. I would frequently have full size hand marks on my back from my brother trying to kill the horseflies that landed there. We could never get them. They always won. Interestingly, both males and females eat nectar and pollen, but females bite because they have to ingest blood before they are able to reproduce. And thus they have been bothering people for hundreds of years. There are references to "gadflies," another name for horseflies, in ancient Greek literature and Shakespearean plays. Loki turns into a gadfly to torment the person making Thor's hammer, Mjölnir. In short, horseflies are an annoying experience shared across time and encapsulated forever in the written word.

When people think of flies, this may be the category they most often put them in. Almost no species are deadly, but some can be described as annoying. They often are the ones who really love what humans have to offer (and throw away), so they may fly around us in excited anticipation or just want to share our yogurt: house flies, horseflies, small flies in small swarms, and botflies. Horseflies cause great pain when you are trying to relax in the water. Large groups of small flies swarming are inadvertently swallowed by people running in the park (even though they are swarming to find ladies). These situations are all preferable to deadly at least. So let's focus on what makes these flies so important.

How Flies Shape Our World

House flies, as mentioned in the intro, are inextricably linked to humans. Without us, they would most likely not survive. For humans, we might survive, but our world would include significantly more trash, and there is a good chance that the food web would start to collapse (a debate among scientists).

The house fly is the quintessential fly. When most people say *fly*, it is what they picture in their mind: stout body, bristly hairs, iridescent or blackish in color, reddish eyes, about a quarter inch, appearing everywhere, often without warning.

House fly feet have sticky, suction cup like pads on the ends of their legs to help them stick to any surface. Their siphoning mouthparts suck up liquid, which is the only form in which they can digest food. They go through complete metamorphosis, so their larval form looks very different from their adult form.

They are so linked to humans that they don't live and thrive in wild areas: they must live near humans.

They do love eating a lot of what we leave behind. House flies and other small flies you see around your domicile or the poop on the sidewalk eat in similar ways. They have what are called siphoning or spongy mouthparts. I know as a kid, if a fly landed on my food, I would refuse to eat it because I'd say that they puked on it. That isn't entirely true or false. They spit a little saliva out onto whatever they want to eat because they are unable to eat solid foods. When it liquifies, they suck it up with their mouthparts acting as a straw. Flies also poop a lot, which is why some people hate them around food: they are considered unclean. Not sure how super clean they are, but if one lands on your hotdog for two seconds, you probably have nothing to worry about. Similar to squirrels, many flies have adapted with humans to eat whatever they eat (and then some), so they love a picnic as much as the next person.

House flies lead fascinating, if brief, lives. They are able to walk on the ceiling because of their tarsi (like little toes) with the help of a small, sticky pad which helps them walk upside down. Even though house flies themselves are pretty slow fliers compared to other flies, their vision is impressive. They can see things at five times the speed of humans, leading to quicker response times. They use their vision for potential predators but also for potential mates. During their short lives, they partake in some acrobatic sex. The male flies into the female midair, and in the fall, he must perform a ritual. If the female is into the ritual, they mate once they land on the ground. If not, she completely ignores him. Males mate with as many females as possible, whereas the females mate with one during their lifetime.

There is a joke in the movie *A Bug's Life* where the main character is at a restaurant and the server yells, "Who ordered the poo-poo platter?" They are then immediately swarmed by flies. When people think of flies, they often think of shit. It's important to note, however, that most adult flies don't eat shit but will lay their maggots in it. House fly and botfly larvae eat poop, but the adult forms eat nectar. Adult house flies are opportunists so will eat whatever they can (including dead material), but they also love dairy and sugary foods.

The aspects that we find gross about maggots and flies are also the aspects that make them ecologically important. Only the flies seem to like the dog poop that your neighbor left on the sidewalk. People find the behavior gross, which is fair because humans don't eat poop and decaying matter. Larvae are the best at decomposing those unwanted materials, whereas most adults eat fruits, plants, and pollen. While some flies thrive on decomposing matter, others are carnivorous. The flies who eat live prey are expert hunters. Regardless of their diet, all flies are an important food sources for other animals.

Despite these gross-to-us habits, flies are not often dangerous, but when reaching large numbers, they have been known to transmit disease. They do not carry the disease within them, but instead have it stuck to body hairs from standing on garbage and poop; however, the random single fly who shows up in your house is unlikely to cause health issues. Still, militaries have been known to use house flies to their advantage in war. Japan used biological warfare on the Chinese during World War II by releasing house flies that had been covered in cholera. Around 450,000 Chinese people died due to diseases introduced by infected insects. Using insects contaminated with disease vectors as biological warfare was banned worldwide in 1972 by the Biological Weapons Convention.

While house flies and horseflies are annoying enough to shape our world in the above ways, botflies are a bit creepier. As in, botfly larvae are like something out of a horror movie. The eggs are laid on the skin of an animal and when they develop, they burrow into the mammalian host. The condition is called *myiasis*. Some species don't even land on their hosts but will shoot maggots at the host from a distance. If you've ever played *Fallout 4*, you've encountered the "bloatflies," which are based off of botflies. In the game, they use their larvae as radiated ammo. Real life botflies are often found in cattle or wild animals but can occur in domestic pets and humans. Chances of being the host in the U.S. are incredibly rare, almost 0 percent. Traveling to warmer, wetter, tropical regions puts you at an increased risk of botfly larvae infestation.

The babies feed and then leave your skin, but it is a fairly unsettling sight. You can have them surgically removed, wait for them to exit on their own, or kill them by suffocation with petroleum jelly (and then have them surgically removed). It leaves a hole in your skin. Nothing good about it, but we can be glad it doesn't kill people. Definitely annoying (and horrifying, and maybe not redeeming) but not deadly, so yay? In wild animals, in large enough numbers, botflies can cause issues in the health of the adults, which leads to issues in the young, especially for those who are feeding from an adult with an infestation. It leads to an overall weakening of the animal in question, but probably won't be the exclusive cause of death.[29]

Despite all that, humans in some cultures have found a way to turn the tables. Botfly larvae have been eaten by humans for centuries, especially during the seasons in which caribou are killed for meat. The Inuit kill the caribou, and when skinning them, collect the botfly larvae that had been living in the animal for food as well. Tit for tat.

How Humans Can Interact with Flies

Flies are drawn to smells that are often considered unpleasant to people. One of those smells is decaying flesh. As a result, they are important in the field of forensic entomology. This practice is when arthropods are used to determine details of the time of someone's or something's death. The use of insects to help solve a crime was first recorded in China in the thirteenth century during a murder investigation: A farmer was killed with a sickle, and all of the local farmers were asked to bring their sickle to the crime scene, and the sickle that attracted flies because of the blood (that was imperceptible to humans) was accused. Sure enough, that farmer admitted to the killing.[30] Forensic entomology wouldn't first be used in a western

---

29 Note: Seriously, do not look up pictures unless you have a strong stomach.

30 Tropiano, D. (2023, Jan 31). CS"fly": Insect data holds clues to crimes. Arizona State University. news.asu.edu/20230130-discoveries-csfly-insect-data-holds-clues-crimes

court of law until the eighteenth century, and then wouldn't become a respected field until about the 1900s.

Forensic entomologists study the kinds of insects found in and near bodies, and by knowing the life cycles of the different insects, and the stages of decomposition when different insects populate dead bodies, can determine the time since death (post-mortem interval), the length of time the body has been there, or if it has been moved since death. Certain species, like blowflies and flesh flies, show up within minutes of death, but arthropod activity continues on the body through the first three weeks after death.[31] K. Tullies and Lee Goff, two entomologists, broke the decomposition process into five stages: fresh stage (day 1-2), bloated stage (days 2-7), decay stage (days 5-13), post-decay stage (days 10-23), and remains stage (after day 18). Flies are particularly active on the body during the fresh, bloated, and decay stages.[32] There are a few ways that this helps to give information about the body. First, certain species show up at different times, which can give the approximate time since death. Second, certain species are only associated with certain habitats so the presence of a species not in its normal environment can suggest the body was moved and where it was moved from. Third, since the flies are eating the body, they can take in any drugs that were ingested by the deceased being, offering more information surrounding their death.

Apparently, some flies also enjoy eating fresh semen. In *Gory Details* by Erika Engelhaupt, a science journalist and editor, she interviews an entomologist who studies the topic. The love certain flies have for semen could have serious criminal repercussions. Say a fly were to feast on fresh semen, fly to another house where a crime happens, and is then killed at the scene. The DNA of the person whose semen they ate would be at the crime scene, even though that person was never there.[33] Clean up after yourself just to be safe.

Maggots of another fly (the common green bottle fly, *Phaenicia sericata*) are used in medicine. These larvae clear necrotic tissue (which is another way to say dead tissue) from a wound and leave the living

31 Joseph, I., Mathew, D. G., Sathyan, P., & Vargheese, G. (2011). The use of insects in forensic investigations: An overview on the scope of forensic entomology. *Journal of forensic dental sciences*, 3(2), 89–91. doi.org/10.4103/0975-1475.92154

32 Tullis, K., & Goff, M. L. (1987). Arthropod succession in exposed carrion in a tropical rainforest on O'ahu Island, Hawai'i. *Journal of Medical Entomology*, 24(3), 332-339.

33 Engelhaupt, E. (2021). *Gory details*. Penguin Random House.

tissue. Certain types of wounds like abscesses, burns, and gangrene respond well to live insect therapy. It greatly reduces the need for amputations. Wounded soldiers on battlefields during the American Civil War were found to be less likely to die from infection if they had had flies lay eggs in their wounds. The little larvae will only eat the dead tissue of the wound, not the healthy. They eat out all of the stuff in a wound that can cause more severe health issues. To them, it is the sustenance they need to continue growing. Many people who have received the treatment say that they don't feel much and that it isn't painful. In the early 1900s, hospitals moved away from the use of maggots in favor of antibiotics; however, since antibiotic resistance is on the rise, many hospitals are returning to live insect therapy. There are a few companies who specialize in providing sterile maggots.

How Human Culture Is Shaped by Flies

Since flies are around us all of the time, it is no wonder that they show up in our culture, sometimes in small, perhaps unexpected ways. In the sixteenth and seventeenth century, it was common to include a house fly in a painting, often in portraits or still life paintings: *musca depicta*, "painted fly" in Latin. Historians have tried to determine the reasoning behind the trend, but no one seems to agree. I am clearly not a historian, but I like the idea that it was included because they are everywhere, often go unnoticed, and provide a sense of imperfection when finally noticed.

In a similar vein, the famous movie *The Fly*, both the 1950s original and the Cronenberg 80s remake, include a fly in a science experiment gone awry. A house fly went unnoticed by the scientists, resulting in their DNA becoming intertwined.

In the world of art, Salvador Dali is known for including flies (and ants) in his works. Some art scholars describe his inclusion of insects as having "nightmare effects". Beelzebub (the Lord of the Flies) is also called Satan (Christianity), a demon (Judaism), and one of the seven princes of hell (demonology). They all agree that he is not a great dude, who is associated with gluttony and envy. Flies do swarm when there is a meal to be had and come right back if you swat them away, which can be anthropomorphized as gluttonous. The bible also isn't doing flies any favors since they are one of the ten plagues. The species of fly that is being referenced in the bible is unclear, but their actions are not: they ruin the land.

Or take poet Emily Dickinson, who loved invertebrates and wrote the famous poem "I heard a fly buzz– when I died". In English, one can use the phrase "a fly on the wall". Similar to *musca depicta*, the phrase "a fly on the wall" shows that a fly can go unnoticed in a room, overhearing everything. Many people attribute the first written use of this idiom to February 1921's issue of the *Oakland Tribune*, "I'd just love to be a fly on the wall when the Right Man comes along." The meaning behind the phrase is that you would love to be able to be in the room and overhear what everyone was saying but have no one notice (like the ever-present but rarely noticed fly). I found the phrase loosely being used in the *New York Herald* from 1889 in a section titled "Look Carefully to Your Judges": "A shrewd reporter is not barred from getting the news from closed doors. Like a fly on the wall, he is always inside of the secret council hall and jots down a note or two as the discussion proceeds." One can talk about a "fly in the ointment" or someone who is a "gadfly" (horse/biting fly).

Moving from fly's influence on language into their impact on film: In 1902, Segundo de Chomon came up with the idea for stop motion animation after watching a fly walk across a sheet of paper, so that's pretty cool. Thanks to flies for animated movies and TV.

Quick Facts: Annoying Flies

- Even annoying flies are important to human culture and the food web.
- Forensic entomologists use flies to determine time of death and other death-related issues.
- Artists from Salvador Dali to Emily Dickinson have been inspired by flies.
- Maggots can be really helpful in hospital settings by eating necrotic tissue.
- House flies are so tied to humans that entomologists believe they could not survive without us.

Bugs Spotlight: Fruit Fly (*Drosophila melanogaster*)

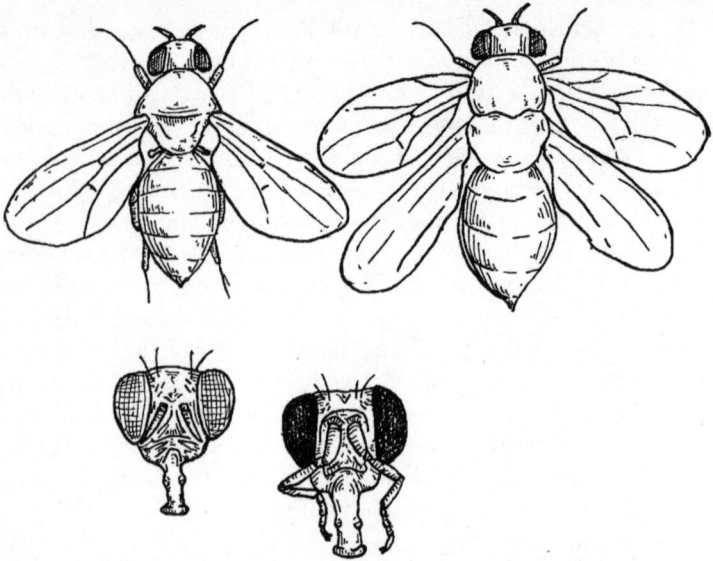

Some genetic experimentation is strange. Scientists were able to give fruit flies a second part of wings (top) and a pair of legs instead of antennae (bottom). These experiments may not be practical but they helped scientists determine which genes in their DNA affected growth of certain features. Seems a bit creepy though.

- Yes, fruit flies can be annoying but are actually harmless.
- Without them, genetic research would not have been able to progress like it did over the last seventy years.
- They are used in research because they have many generations in a year (one generation about every ten days) and females lay upwards of 2,000 eggs in a lifetime (which is only about two weeks each).
- They are attracted to decomposing food so if you do not want them in your house, be sure that your food scraps are kept outside and your fruit isn't going bad.
- *Drosophila* is Latin for "dew-lover."

Chapter 6: The Potentially Deadly Flies

Unfortunately, two types of flies can cause serious illness in humans and other animals. Mosquitos, who are dipterans, cause more human deaths per year than any other animal (apart from humans) due to transmitting diseases like malaria, dengue, Zika, and West Nile virus. Tsetse flies are known for their transference of sleeping sickness to humans and animals (called *Nagana* in non-human animals).

To this day, these creatures thwart use of certain land. Currently, parts of Africa are overrun by tsetse flies and people cannot move into or grow food in those areas. In the 1500s, mosquitos delayed the Spanish invasion of the Yucatan peninsula (for a while at least).

Up until recently, people in the U.S. did not often have to deal with disease spread by dipterans, at least not within this century. With global warming and increasingly wet conditions, these diseases are making their way into the U.S. And so scientists are working on ways to reduce the impact of introduced illnesses.

How Flies Shape Our World

Mosquitos are not who you think of when you think of deadly animals. People pick things like sharks and lions, snakes and bears—not a very small insect.

In the summertime, people all over the U.S. know when it is mosquito season. These bugs swarm and bite at various times of day, so you never know when they're coming. In backyards, people spray bug spray, wear long sleeves and pants, light citronella candles, and have mosquito traps to deter being bitten. Even with all of these preventative actions, you are still likely to get a bite or two. Then you have to deal with the itchy lump for a few days. In areas where numbers are particularly great, people use mosquito nets and window screens treated with insecticide. And so we all curse the mosquitos and question why they are here and if they are really needed.

That was even more so the case during the 2015 scare of Zika virus that made its way into the U.S. Zika is a virus that often has few or sometimes no symptoms. Most adults never realize they have it. The reason it was such a scare was that it can severely affect the fetus in pregnant women. The results include microcephaly and other severe brain malformations. Microcephaly is when the head develops smaller than normal, leading to issues with brain development. The disease was normally only found in a small portion of Africa and Asia but was spread to the Americas. The U.S. government warned about travel to where the virus was known to be prevalent. It can also be spread sexually, most often from men who have it to their partners. As of 2024, there is no vaccine for the virus.

Malaria is a disease that infected 247 million people in 2021 and caused 619,000 deaths that same year.[34] Most of the deaths were in sub-Saharan Africa (95 percent) and were in children under the age of 5 (85 percent). The U.S. has been malaria free since 1970, although people who travel to areas with malaria have brought it back with them.

Eradicating mosquitoes is not possible, but eradication of diseases is with preventative medicines and vaccines. In 2023, a malaria vaccine was given to children that resulted in 15 percent less deaths in areas with medium to high infection rates. Many countries have created their own malaria eradication goals. Forty-four countries, through their own methods and help from worldwide organizations like WHO, are certified malaria free. Only five countries in Africa are malaria free, the most recent being Cabo Verde in 2024. The havoc this disease wreaks is so closely tied to colonialism, forced poverty, and ongoing economic issues. Layer after layer could be peeled apart in a book all its own.[35]

Dengue, similar to malaria, is most commonly found in the wet forests of Asia and Africa. There are also large numbers of dengue cases reported in Brazil, linked to their rainforests being ideal areas for mosquitoes to breed. Unlike with malaria, there are two vaccines

34 World Health Organization. (2022). World Malaria Report 2022 [Annual]. who.int/teams/global-malaria-programme/reports/world-malaria-report-2022

35 And it has been. Check out these articles for more information:

Bump, J. B., & Aniebo, I. (2022). Colonialism, malaria, and the decolonization of global health. *PLOS global public health*, 2(9), e0000936. doi.org/10.1371/journal.pgph.0000936; Great article about colonialism and malaria; interesting article on social implications of malaria (Heads up: This one is kind of problematic because it is without reference to colonialism); Ricci F. (2012). Social implications of malaria and their relationships with poverty. *Mediterranean journal of hematology and infectious diseases*, 4(1), e2012048. doi.org/10.4084/MJHID.2012.048

available for the fever (one that was just released in 2023) and when adequate treatment is received, the chance of death is less than 1 percent. If the infected person is stable, they can stay home and drink lots of fluids to recover. Many countries still have outbreaks, including one in 2020 in Latin America. Outbreaks are more common in highly urban environments. The continuation and increase of outbreaks is attributed to population growth, international travel, and global warming.[36]

West Nile was first reported in the U.S. in 1999, most likely starting with an infected bird that was then fed on by a mosquito who then drank from a human. This virus is less deadly to humans than horses (about 7 percent fatality in humans versus 40 percent in horses). 7 percent is still no laughing matter. Developing encephalitis or meningitis only happens in about 1 percent of the people with West Nile. For those who do have symptoms, it can take weeks to months to feel better. There are now four different vaccines for horses but none for humans.

The CDC recommends the normal precautions for humans to reduce mosquito bites: long sleeves, insecticide, limiting time outdoors, fixing screens, and removing standing water (which can cause mosquitoes to reach high numbers). News programs may tell you to dump out any standing water in unused tires or birdbaths. But it's important to note that some of the standing water is not controllable by regular people. Global warming is increasing temperatures and severe flooding. These hotter temperatures can lead to higher evolution rates of the virus, and more standing water from flooding can lead to higher numbers of mosquitoes.[37] Both of these factors increase the likelihood that we will see higher incidences of West Nile in the future.

All of these diseases are responsible for lots of pain and suffering. They are often the most problematic in regions that don't have access to vaccines or preventative drugs. Places like the U.S. have vaccines available, as well as government-controlled programs that deal specifically with the eradication of these illnesses. Places like sub-Saharan Africa and the equatorial belt in Asia are not able to access care as easily because of the high cost. Capitalism and colonialism

36 Whitehorn, J., & Farrar, J. (2010). Dengue. British medical bulletin, 95(1), 161-173.

37 Paz S. (2015). Climate change impacts on West Nile virus transmission in a global context. Philosophical transactions of the Royal Society of London. Series B, Biological sciences, 370(1665), 20130561. doi.org/10.1098/rstb.2013.0561

continue to create issues for basic human rights. The World Health Organization is trying to break down those barriers to healthcare, but that dismantling takes time, energy, and resources that are difficult to set in motion.

As for the mosquitoes, they are just doing their thing, unaware of how dangerous they are. Adult female mosquitoes need to feed on vertebrate blood in order to produce young (in most species). They use their stabbing mouthpart to feed off of humans, domestic pets, cattle, and wild animals. Male mosquitoes do not bite or drink blood, but instead drink nectar. When the female bites a vertebrate animal that is carrying one of these diseases, it is transmitted to them but does not cause adverse effects for them. The virus lives in their abdomen, either fully formed or growing there, and is then released into the next body through the mosquito's mouthparts.

Again, mosquitos are not the origin of the virus; they instead transmit it from one source to the next. I'm not sure that the phrase "don't kill the messenger" works in this situation. Scientists debate whether or not we could live without these bugs. Many believe that it would create a collapse in the food web since so many animals rely on them (bats, birds, predatory insects, fish), whereas other scholars believe something else would fill that gap. We can't know for sure, but we have seen before that losing an entire species can have a serious, negative ripple effect on the ecosystem.

How Humans Can Interact with Flies

Tsetse flies are another species of dipteran which has caused serious issues for humans. They pass on sleeping sickness to humans as well as cattle. For those who are affected most severely, those in sub-Saharan Africa, this is a double whammy. Not only are the humans at risk for getting the infection, but their livelihood of cattle is also in danger. Cattle are weakened and most likely will die, females cannot carry their young to term, and then a lack of fertilizer is available for farming other crops. If not treated quickly, sleeping sickness in humans is almost always fatal. There are drugs that are used early in the course of the infection that are lifesaving, but for one strain (which accounts for 98 percent of cases) a person may not have symptoms for months or years until the advanced stages. Once a person has reached that stage, there are no drugs to help. Tsetse flies currently reign over about 4 million square miles of green space that no one can use. For those who do live and work in areas with the flies, they are devastated by the effects.

In an attempt to control mosquitoes and tsetse flies, scientists have continually tested the sterile insect technique (SIT) as an ecologically safe alternative to using insecticides. The process includes sterilizing large numbers of males by using radiation. These males are then released into the local population. They behave normally in every other way and are unaware of their sterility. This leads to them "fertilizing" eggs and courting females, which in turn leads to a drastic decrease in larval numbers. For tsetse flies, the results are very promising. In Zanzibar and Senegal, the population was eradicated or reduced by 99 percent. Unfortunately, while this application has been successful in the laboratory and in smaller, controlled studies, the process can be prohibitively expensive and may not work in areas where multiple species transmit the disease.[38]

For mosquitoes, lab-based studies and small scale field studies have shown promise. Three species of mosquitoes are responsible for most of the serious diseases and infections. All three currently target species in many SIT studies. In 2021, researchers did a study in urban areas of Havana, Cuba, which resulted in a suppression of the mosquito population.[39] Work will continue to be done to determine how plausible, both financially and scientifically, such a practice would be at a larger scale. This practice would also leave the insect populations stable, allowing for their continued benefit as an important food source, without devastating human populations. Everyone would win.

How Human Culture Is Shaped by Flies

The U.S. military during the Cold War tried to determine if weaponization of mosquitos was worth it. They had numerous operations: Operation Big Itch, Operation Big Buzz, Operation Magic Sword, Operation Drop Kick, Operation May Day. All of these operations released uninfected mosquito/flea tests to determine if the flies could survive being dropped out of planes from high/far

38 Brun, R., Blum, J., Chappuis, F., & Burri, C. (2010). Human african trypanosomiasis. The Lancet, 375(9709), 148-159.

39 Gato, R., Menéndez, Z., Prieto, E., Argilés, R., Rodríguez, M., Baldoquín, W., Hernández, Y., Pérez, D., Anaya, J., Fuentes, I., Lorenzo, C., González, K., Campo, Y., & Bouyer, J. (2021). Sterile Insect Technique: Successful Suppression of an Aedes aegypti Field Population in Cuba. Insects, 12(5), 469. doi.org/10.3390/insects12050469

distances. They informed the cities in the U.S. they were going to test them on. The tests determined that mosquitoes released would be able to survive and spread disease effectively. The military states that they never followed through with using these methods in war.

Quick Facts: Deadly Flies

- Only two species are deadly to humans: tsetse flies and mosquitos. Mosquitos have caused more death than any other animal.
- Programs releasing sterilized males have been successful in reducing mosquito numbers.
- The U.S. government tried to see if weaponizing mosquitos was possible. It technically was, but they claim they never used them in that capacity.
- Tsetse flies are responsible for around 4 million acres of green space being inaccessible to the people of Africa.
- A few countries have successfully eradicated tsetse flies, but quite a few countries are still plagued with them.

Bug Spotlight: Mosquito

A male mosquito eating nectar at a flower and a female hovering. You can see from the top-down view that they only have two wings, like all flies.

- Mosquitos may be what took down Alexander the Great.
- The Panama Canal was built and then controlled by U.S. after earlier attempts were thwarted by mosquito-borne yellow fever (by the French).

- Programs releasing sterile males have been fairly successful in reducing numbers.
- Adults are terrestrial while larvae is aquatic.
- Many species have no interaction with humans.
- In *Jurassic Park*, mosquitoes in amber were used to make dinos, which probably wouldn't be possible.

Sorry for such a downer of a chapter. It is interesting to know that out of the 150,000 species of flies, only about ten cause serious issues for humans. They affect life in such a way that it would be irresponsible to not discuss it, even if it paints them in a negative light.

Part III The Most Plentiful: Beetles (Order Coleoptera)

One in every four animals in the world is a beetle. One in every ten animals is a weevil (a type of beetle). They are the largest order of insects with about 400,000 known species. The word *beetle* comes from the Old English verb "to bite" even though many of them don't bite, including the ones with big chompers like stag beetles.

Their scientific name is Coleoptera which means "sheath wing" or "hard wing." Beetles have a hard outer wing and a soft (often foldable) pair of wings underneath. Think of a ladybug: their red, dotted shell is their first pair of wings while their second pair hides, folded beneath that.

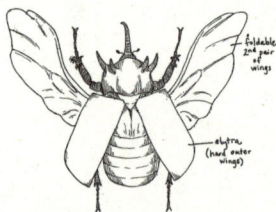

The five horned rhinoceros beetle, *Eupatorus gracilicornis*, with clearly defined hard outer wings (elytra) and softer, foldable under wings. These beetles can get up to 3.5 inches in length.

Their first appearance in the fossil record is from about 295 million years ago in the Permian period. They survived two mass extinctions but are, like other insects, not doing well overall in our current, human-caused sixth extinction. We are in the age of the anthropocene, which is the first time in the world's history that humans have been the dominant influence on climate and the environment. Our practices have led to a decline in biodiversity and animal populations. Beetles are no different.

Since there are so many species, it can be hard to generalize their behaviors or to know which species is being specifically referenced in history and folklore. The ancient Greeks believed that all beetles sprung from and lived in fungi, which is only half correct. Some species of beetles, called fungus beetles, are often found on and around fungi but are not born of it. Other references are very clearly a specific beetle, at least a specific family, like scarab beetles in ancient Egyptian culture. Egyptians viewed this bug's poop-ball rolling to be similar to the sun rising and setting. Their sun god Khepri has the head of a dung beetle.

Beetles that feed on plants can help with pollination using their "mess and soil" method. Many beetles have fuzzy thoraxes and abdomens that pollen can cling to. Species of click beetles, scarab beetles, longhorn beetles, and a few others can potentially pollinate. These evolutions may connect to plants they are seen on most often. Some insects rely on specific plants for food. The longhorn milkweed beetle eats milkweed in both its larval and adult form. They use the poison from the plant as one method of defense, and also use their red and black dotted body to warn would-be predators about their poisonousness. Some beetles are considered pests, and most of the current pesticide use on farm land is for beetles, mostly weevils.

Beetles' defenses range greatly. When disturbed or attacked, most beetles have a "drop off reflex," or closure method, where they bring their legs close to their body or go limp. Lady bugs just play dead as a defense mechanism. Many beetles have warning colors that are their initial line of defense. Another form of defense is using mimicry or camouflage. Some beetles have markings to trick predators into thinking that they have a stinger (even though they don't). Some are colored similarly to their surroundings, like leaves or sand, so they don't stand out: when they are stationary, it's almost impossible to see them.

As I mentioned above, it is hard to group beetles together to generalize behavior. The suborder Polyphaga contains the largest number of species, including some of the most well-known types of beetles. The other suborder Adephaga contains the beetles that people probably most often see in urban areas (ground beetles) but some others that are pretty rare (and becoming less and less sighted due to dwindling numbers). There are so many beetles that won't get deserved attention here, but know that they are also important in their own spheres. The two groups I discuss here account for about 390,000 of the known 400,000 species.

CHAPTER 7: LADYBUGS, SCARABS, AND WEEVILS (SUBORDER POLYPHAGA)

The variety of beetles in the suborder Polyphaga is astounding. Of the 400,000 known species, this suborder includes 90 percent, around 350,000. The outward appearance of members of this suborder can be drastically different: compare a ladybug to the longhorn Palo Verde beetle. Their differences are great in almost every way: size, shape, food source, public opinion. You may be wondering how they could even be related. Most beetles have two sets of wings, with one hard set (called elytra) and a set of softer, often folded wings underneath.

Similar to butterflies and moths, beetles go through complete metamorphosis similar to butterflies and moths. The larval form of most beetles (what are commonly called grubs) are very different from their adult form. Unlike butterflies and moths, the larvae of beetles are pretty immobile and eat underground—mostly stationary to reach the next stage in their development.

Larvae size can often give you an indication of how big the beetle will be once it reaches adulthood. The hercules beetle is the largest and heaviest living beetle in the world. They live in Central and South America, so you won't see one in the backyards of the U.S. As adults, males (with their long horn on the front) can reach seven inches. Their horn plays an important role in mating, which I'll talk about more later. Females do not have the horn and can still get to about three inches. The grub can get to about four and a half inches and weighs a quarter of a pound. Hercules beetles are also able to comfortably carry around fifty times their body weight, making their strength comparable to ants.

The largest beetle that is found in the U.S. is a relative of the hercules beetle, called the eastern hercules beetle. It is not quite as big as its cousin and the horns on males are apparent but shorter. Everywhere I looked mentioned the eastern hercules beetle and the giant stag beetle; but nowhere did I see a mention of the Palo Verde beetle, which is native to Arizona. Let me tell you, those bois are huge, and I have no idea how they aren't on any of the biggest beetles in the U.S. lists. The first time I saw one, I was genuinely shocked by their size. They can get up to three inches and have long antennae as well. Root borers in general can get pretty big.

Apart from size, beetles have other interesting attributes. Sexual dimorphism is the term for when males and females of the same species look extremely different. As I mentioned about the hercules beetle, the males have long horns whereas the females don't. This is the same for the eastern hercules beetle: horned males, hornless females. The Guiness Book of Records surprisingly has a category for the greatest sexual dimorphism, which is held by the trilobite beetle (a member of the Polyphaga family). Both the males and females of the species look non-beetle-like. The adult males can have softer elytra that don't connect at the bottom, giving them an almost moth-like appearance. Females look like the now-extinct trilobite fossils.

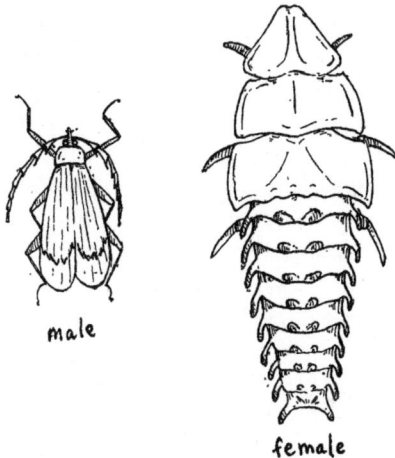

male

female

Sexual dimorphism is common in insects. The trilobite beetle males and females look like different species. The female has no wings and resembles an ancient arthropod. The males have soft elytra and fly to find females. Females are about 3 inches in length whereas males are only .3 inches.

Another fun distinction for a beetle is a South African weevil with a big schnoz. This weevil is only about a centimeter long but it has a nose that adds two centimeters. It has a nose that is twice the length of its body for females, whereas males have shorter snouts. The use of the nose is to "drill" holes in plants to lay eggs.

How Ladybugs, Scarabs, and Weevils Shape Our World

Only 1 percent of beetles are defined as pests. The largest group is weevils (which is also the largest group of beetles). The weevils that are pests are often named after a food that they eat: rice weevil, granary weevil, corn weevil. Pest management is often targeted towards them to keep up with their ability to evolve when confronted with new pesticides. The boll weevil is considered to be one of the worst weevil pests in U.S. history. It is estimated to have caused $16 billion in losses throughout its time in the U.S. It was first identified in the U.S. in 1854. These bugs migrate via wind so they just floated over here from further south. In the 1940s, the weevil decimated the cotton crops of the south, particularly in Alabama. Bugs need food and are likely to eat what is available to them. Sometimes, they like (or need) certain plants over others. The way farming is done today increases the likelihood that crops will have pest issues like weevils. Why is that the case?

If you only ever plant one type of plant, like cotton, and there are no other plants nor are there any natural predators of the insect (due to accidental introduction), you are in for a bad time. The insects have enough food to reproduce and feed their young, who then reproduce at least a couple times a season. Without predators to control the population through natural means, the pests can reproduce exponentially. The bugs will win in that situation every time. Alabama changed their entire industry to include other crops like peanuts because of the destruction weevils caused. In downtown Enterprise, Alabama there is a weevil statue to remind the community of the difficulties they overcame and the importance of avoiding monoculture farming. Again, planting monocultures (just one type of crop) makes it easier for pests to cause issues.

Another well-known beetle pest is the Colorado potato beetle. Countries have done experiments to see if they could be used for biological warfare. They are known to be such voracious eaters that they can decimate crops. The U.S. and other allies were trying to determine if dropping them on Germany would cause food issues during the war. France, Germany, and the United Kingdom were testing whether or not Colorado potato beetles would be a great use of ecological warfare in World War II. The U.S. sent 15,000 beetles to the UK for study. The Germans also used 54,000 in an experiment. At some point, some escaped the lab. Now, the non-native beetle has established itself in Europe. No one claims responsibility for the introduction.

Two other pest species that come to mind are the emerald ash borer and Japanese beetle. Both species were introduced by humans. The emerald ash borer was first identified in Michigan in 2002, though scientists believe it was in the U.S. as early as 1996. In its native China, these bugs are not considered pests and only go for trees that are ailing. In the U.S., they love healthy ash trees and are responsible for killing millions of them within the first decade of their arrival. The government invested in eradication efforts, but the beetles continued to spread: they imported two parasitoid species of wasp to try to control populations, but it will take time to determine if they are making an impact. Cities have programs to check and verify ash trees as borer free through preemptive measures; however, they do continue to spread and people are concerned for the future of ash trees in the U.S.

In Japan, the Japanese beetle is not a pest because it has natural predators and the foliage is less attractive to them. In the U.S., people get really pissed about these bugs. They eat many different plants, around 307 species, but they are particularly annoying to people with rose bushes or vines. The Xerces Society has an interesting training on garden pests where they discuss issues with insects like Japanese beetles.[40] They say that bugs need to eat and that some destruction is probably okay. If they are eating some of the plants in your garden but most of your garden is doing well, is that actually a problem? It's about balance and is something worth thinking about.

Other beetles are considered household pests. Carpet beetle adults love flowers and nectar. Carpet beetle larvae on the other hand love munching on natural fibers: they are known to reduce pinned insect collections to powder because they love eating exoskeletons. Hair, both human and pet, is also an attractive snack for them. I frequent the bug identification subreddit and the most commonly asked identification question is about carpet beetles, so much so that they pinned a picture at the top saying, "Hey, is this what you saw? It's a carpet beetle" so that people would post about it less.

---

40 You can check out the video addressing pests in your garden at youtube.com/watch?v=1eAuQOnGJlQ.

actual size

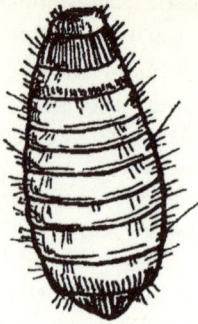

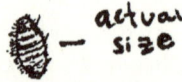

Carpet beetles are tiny, but the larvae like to eat natural materials so they can often be found amongst pet hair, on your bed eating sheets or dandruff, or on your bookshelf. A nuisance more than a serious problem. Adults eat nectar but lay eggs in natural materials.

Many people fear that they are bed bugs, but they do not bite and are much easier to get rid of. They aren't in the same order either. In large numbers, however, carpet beetles can do some damage to your clothes or comic books but are not impossible to get rid of. They can be eliminated with regular, thorough vacuuming.

Ladybugs are the opposite of a pest. They, particularly the larvae, are known to be avid hunters of aphids. Ladybug larvae can eat up to seventy-five aphids a day. Integrated Pest Management, or IPM, is an ecologically-friendly form of pest management: less chemicals are used on produce, natural invertebrate enemies are used for human gain, and less issues with groundwater and crop destruction. Some pest management companies rent out ladybugs. Scientists, however, warn against people buying ladybugs to put in their own garden. Since adults can fly, there is a good chance that the ladybugs you buy will not stay in your garden. There is no guarantee that they will stay on your plants. Asian lady beetles (another name for ladybugs) were introduced to the U.S. for pest management purposes but are now established all across North America. Fun fact: ladybugs are poisonous if eaten. Their colors indicate that. They also have rampant STDs.

Ladybugs are known to transfer both a sexually transmitted fungus and mites. Neither immediately kills the host, but they will affect the overall health of the insect. Apart from the diseases ladybug's transmit to each other, one wasp species uses them to rear their young: the green-eyed wasp lays an egg in the ladybug. When it hatches inside the body, the ladybug goes about its business. The larva eats its organs, leaving only the organs that are needed to keep the ladybug alive. When the larva is ready to emerge, it turns the ladybug into a zombie. The ladybug becomes paralyzed while the larva emerges from its body. The wasp larva spins a cocoon to grow in, wrapping around the ladybug's legs so that they can stand guard.[41] The larva pupates and then flies away. Sometimes, the ladybug lives and goes about their life with fewer organs and a hole in their body.

How Humans Can Interact with Ladybugs, Scarabs, and Weevils

When people think of other beloved insects, fireflies often come to mind. Many people in the Midwest and Eastern U.S. have childhood memories of chasing lightning bugs during the summer. Some people may have noticed smaller numbers over the last few decades. Why might that be the case? Entomologists believe fireflies are like miner's canaries with respect to several ecological concerns. They are sensitive to pollutants, human activity, lights, and pesticides. Scientists believe that if fireflies are disappearing from an area, we need to research those four human-caused issues to determine if the habitat is in more danger than previously thought. Several undescribed and endemic species discovered in north central Florida forty years ago have not been seen for several years. Anthropogenic (human-caused) factors are the most likely culprits. You can help by turning off your lights at night, like outside lights you may not actually need—to 1) not confuse them when they are looking for a mate and 2) reduce the lighting so that they can see females. Both males and females have glowing butts. Most male flash patterns are species specific. They are the ones you see flying around. Females respond with a single flash while clinging to the end of a blade of grass. Some females mimic the flash pattern of females from other species to lure and eat males. Males greatly outnumber females during mating season so eating a few isn't devastating to overall numbers. They taste terrible as a defense mechanism to deter animals from eating them. They are even deadly to some animals.

Some beetles have co-evolved with plants and have formed essential relationships for pollination. Magnolia blooms are best pollinated by the tumbling flower beetle. Many insects visit those flowers, but the

41 Weiler, N. (2015, Feb 10). *Wasp virus turns ladybugs into zombie babysitters.* Science. science. org/content/article/wasp-virus-turns-ladybugs-zombie-babysitters

tumbling flower beetle is the first to arrive in the season and is also the most efficient pollinators of that flower. Their style is called "mess and soil," similar to butterflies and moths. They rummage around flowers, feeding on pollen, which is then dragged to another flower on their legs/abdomen. Clumsy but effective. They are attracted to smells that are slightly stinky, similar to decomposing plant life (kind of like flies). Many beetles are detritivores, meaning they only eat decomposing plants, so naturally they'd be attracted to those smells.

Dung beetles, while not considered the cutest of beetles (because of the poop), are harmless and important. They may not have cute spots or flashy butts, but they do help with nutrient cycling, waste removal, and the suppression of various flies/parasites. Unfortunately, several studies show a decline in biodiversity and in abundance of dung beetles with loss of habitats as a result of building/coastal development. It also doesn't help that they have low fecundity, meaning they don't lay many eggs. On average, each pair has about twenty offspring per year, but in some species, it can be as low as five. In cattle pastures, the current diversity of the species in the U.S. is low and many of the species are small-bodied. In healthy populations, upwards of 400 beetles can be attracted to and feast on elephant poop in 15 minutes. They may use dung for 1–4 weeks, depending on moisture. They make their ball of dung and lay one egg per ball. Both parents take extensive care of the eggs and the young when they hatch. Most males will abandon the nest if another male is found fooling around with the Mrs. If caught cheating, there is a chance that the young ones won't have biparental care.

Dung beetles roll poop backwards with their hind legs. Sometimes the poop balls can be significantly larger than the beetle. Leg day every day.

Overall among beetle species, parental care is uncommon, but a few of them, like the dung beetles, provide extended childcare. Burying beetles will prepare food (i.e., dead animal carcass) for their larvae. The males stick around through half of the growth, but females stay the entire length to adulthood. They may even search for food with their adult children. Burying beetles are important to the decomposition of dead animals, which makes them a part of forensic entomology. While flies arrive immediately to the scene and stick around for most of the decomposition process, burying beetles can be found on bodies later in the process. For smaller animals, they bury the body for their young, hence the name. Their young eat the carrion while the adults eat the maggots of flies that are also around the body. The American burying beetle, which was once common, is considered endangered due to habitat destruction.

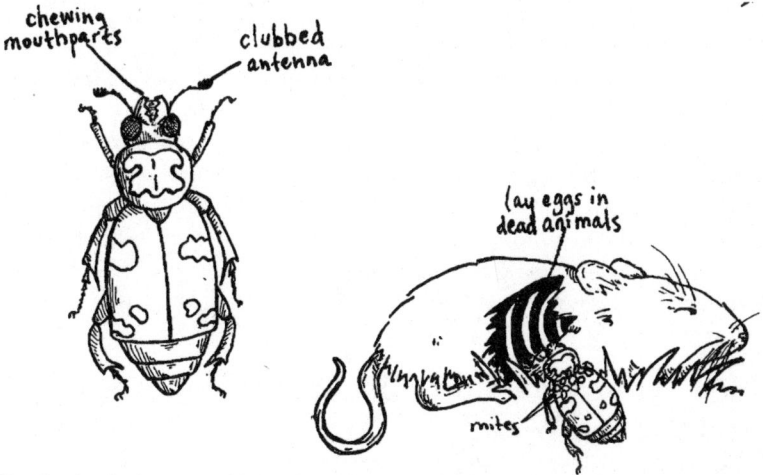

Burying beetles have a symbiotic relationship with mites. You will often see large numbers riding on the back or legs of the beetle, like in the illustration on the right.

The Saint Louis Zoo has an exhibit called the Insectarium (funnily sponsored by Monsanto, a pesticide company). The biologists who work with the zoo started a conservation project to address issues surrounding the endangered American burying beetle in Missouri. In collaboration with the Missouri Department of Conservation, U.S. Fish and Wildlife, Missouri Department of Natural Resources, and the Nature Conservancy, the Saint Louis Zoo has been part of a breeding program. They started the project by determining the overall state of the beetle in Missouri. At one time, the beetles were found in 35 states but as of 1989, they were only recorded in a single state.[42]

The decline in numbers is believed to be habitat loss and fragmentation (there is an obvious pattern here). Fragmentation is when areas that insects live in are cut off from one another, by a street or new development.

Fragmentation happens with the spread of humans into different areas. What starts as an open field is separated into much smaller patches of land, making it hard to traverse from one patch to the next, as well as offering less food and area for breeding/living.

Fragmentation makes moving from place to place difficult and dangerous, often resulting in bug's inability to spread out and reach places they once could. In 2012, the Saint Louis Zoo released hundreds of beetles to places around Missouri. The beetles had been bred and reared in captivity to be released. They are continuing to monitor the situation and determine if their breeding program is helping. Fingers crossed that this work increases population numbers. The role of beetles as decomposers in the environment is essential to a healthy ecosystem.

How Human Culture Is Shaped by Ladybugs, Scarabs, and Weevils

The world of beetles includes interesting mating rituals. Hercules beetles (named after the Greek god) have horns that are on the top and bottom of their heads, not their mouthparts, and they try to get the other male's horns in between their horns so that they can throw them off of the branch they're on. Male stag beetles, the ones with large pincers, are known to fight for the female. In Thailand, people gamble on fights where they place a female in the "ring" and the pheromones she releases will cause the males to fight.

42 Center for American Burying Beetle Conservation. (n.d.). *Background*. St. Louis Zoo. stlzoo.org/conservation/in-action/saint-louis-wildcare-institute/center-for-american-burying-beetle-conservation-2.

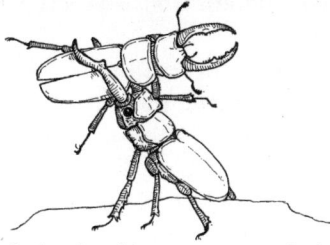

One stag beetle suplexing the other, a common practice in stag beetle fighting.

The point of their mandibles is for that purpose only, similar to deer antlers. Larger mandibles mean a more virile male. Even with all of this fighting, males and females have sex with multiple partners regardless, so the fighting is only somewhat important.

Some people really enjoy the mandibles of stag beetles but instead of making them fight, they keep them as pets. In Japan, stag beetles used to be available in vending machines. The machines were so popular that they would quickly sell out. Most of the same machines today only carry plastic versions (which is better from an animal welfare perspective). They are said to bring good luck to whoever owns or carries one. It was a fashionable accessory at one point in time.

In fact, there is a long history of people using beetles in fashion. A rumor about decorated ironclad beetles from Mexico made its way to the U.S. Historians can only find evidence that it was popular in the U.S.A. (not Mexico) in the Victorian period. The idea was that an ironclad beetle, which can live for a few years and is fairly slow moving, would be decorated with rare materials like gems and gold. Then, the beetles were attached to a woman's collar with a small chain. The beetles crawled around their shoulders all evening. Women were also known to put fireflies in their hair or on their dresses during that time period. Scholars speculate that this practice was popular because interest in natural history was at its height.

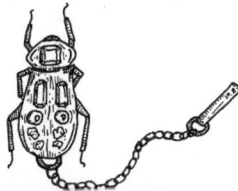

The chain is glued to the underside of the beetle, which then keeps it from traveling too far on your shirt. Some people regard this practice as cruel.

Cities in Mexico do sell these ironclad beetles covered in gems but mostly just for tourists.

In some Asian countries, like Thailand, Myanmar, and India, a similar practice exists (i.e., only for tourists). At one time, beetle wing art and fashion were extremely popular and considered an ancient art. It is not practiced as much anymore, but some people still produce earrings and simple designs for tourists. In this practice, artists would use the hard wings (elytra) of metallic borer beetles to create intricate patterns on canvases, clothing, and jewelry. I used to have a pair of earrings that were made with metallic borer beetle wings, and I've seen quite a few people wearing them. They are very light and make a ridiculous amount of noise in the wind. I can see why using them in art or clothing would be better. Helps the wings be more secure and less likely to bang around.

In addition to uses in fashion, beetles show up in comics, the small and big screens, and music. I'm sure you have heard of The Beatles, a popular band from the 60s. They chose their name (with a different spelling) because, as George Harrison said, "We were just racking our brains and John came up with the name Beatle. It was good because it was the insect and it was also a pun, you know, 'beat,' on the beat." Another beetle recently had a movie: *Blue Beetle* (2023) from the DC comics. Apart from his name and his ability to fly, Blue Beetle doesn't share many similarities to beetles. A large scarab is what gives the character his powers.

Scarab beetles in particular are common insects in stories, movies, and other forms of popular culture. Egyptian culture revered the scarab beetle (or dung beetle), so it appears in films like *The Mummy* or TV shows like *Moon Knight* (Marvel). The representation of the beetles in *The Mummy* is highly inaccurate. Scarab beetles do not burrow under your skin to eat your insides. I know that a lot of people were grossed out by that and worried about something similar happening to them. Impossible. Their little mouthparts could not break the skin and they have no interest in eating human insides. *The Mummy*, however, would be a great movie to show at the Insect Fear Film Festivals (more on that in chapter 11). In *Harry Potter*, Rita Skeeter gets turned into a beetle and is kept prisoner by a high schooler in a jar, a situation which is much more messed up than I thought it was as a kid.[43]

43 NOTE: Too bad *Harry Potter* is such a bummer since JK is a piece of shit. Dung beetles wouldn't touch her though.

Quick Facts: Ladybugs/Scarabs/Weevils

- Beetles are the largest number of insects in the world.
- 1 in every 10 animals in the world (not just insects) is a weevil.
- Weevils are responsible for the Southern U.S. diversifying crops and moving away from only growing cotton.
- Some beetles are great pollinators, like the tumbling flower beetle and the magnolia tree.
- Fragmentation and habitat destruction are leading to reduction in insect numbers, like burying beetles. Certain organizations like the Saint Louis Zoo have projects that are focused on increasing the numbers of beetle species.

Bug Spotlight: Scarab Beetle

Side view of male rainbow scarab beetle (top left); top view of female rainbow scarab beetle (notice the lack of horn) (top right); a common necklace found in ancient Egypt of a scarab beetle carved in stone (bottom)

- The rainbow scarab beetle is a type that is only native to the U.S., who loves Bison poop.
- They were extremely important in Egyptian culture and were represented in writing and jewelry.
- Some are very attentive parents and make sure their babies, of which they may only have a few a year, grow up successfully.
- Loss of food, habitat destruction, and fragmentation threaten the future of scarab beetles.
- Scarab beetles can take down an entire elephant poop patty in fifteen minutes and then manage to use it for upwards of four weeks.

Chapter 8: Ground, Tiger, and Diving Beetles (Suborder Adephaga)

The suborder Adephaga is second in terms of size after the suborder Polyphaga. Fossil evidence of this suborder goes back 250 million years. Ground beetles account for around 40,000 species in this suborder. Ground beetle is a fairly loose-sounding term, and many of the species have significant differences. In general, they are black or brown, but some species can be iridescent (like the caterpillar hunter) or have colorful heads (like bombardier beetles). Many species have defenses, like hot liquid or foul smells, to deter predators.

In this vein, I have a vivid memory of being in the car with my dad, windows down. Something huge flew in through the window but we couldn't see what it was in the dark of night. We noticed an unpleasant odor, but still no sign of the bug. I looked up and saw that it was crawling on my dad's shirt collar. He felt it as soon as I saw it and tried to throw it out the window. He missed, and the car smelled even worse. When we got home, I was able to grab it off the floor and look at it before releasing it. It was a gorgeous iridescent caterpillar hunter.

These bugs get their names for being effective hunters of caterpillars. Many of the species in this suborder are considered beneficial insects. The caterpillar hunter is, of course, a predator of some known lepidopteran pests. They do their part in trying to keep the numbers of those pests down. Ground beetles in general are often predators of other insects, making them an important group for ecological stability. Other ground beetles are carnivorous, but some eat plants or decaying matter. Many beetles can eat their weight in food each day. I can't imagine eating 160 pounds of food every day.

Ground beetles are probably the beetles most often spotted in urban areas. They are incapable of stinging, help control pest populations, are important for decomposition, and are an integral part of the ecological web. They can move pretty quickly, which may be shocking; however, they are harmless and not a terrible thing to find in your house. If you do find them there, they are probably lost and just trying to get somewhere else. They don't live in human dwellings or find them ideal.

How Ground, Tiger, and Diving Beetles Shape Our World

Most ground beetles, like your common ground beetle who is blackish or brownish, do not have defense systems. They just flip over and tuck in their legs in the hope that you'll leave them alone, essentially playing dead. It is fairly effective without much effort on their part. However, some species, when disturbed, may use their other defense systems. A few will release a reddish liquid from their legs that is caustic or smells bad. They will play dead and release the liquid. It reminds me of horned toad lizards shooting blood at predators from their eyes. A little less squirty in bugs—but the same idea.

Some of the evolutionary behaviors of beetles almost sound fake. Bombardier beetles are capable of shooting a blazing hot liquid at would-be predators, at a temperature of close to 200 degrees. Unsurprisingly, this is deadly to some predators and can cause burns on humans. They actively hunt at night and you are not likely to run into one, so no worries. Similar to unknown caterpillars, it may be a good idea to not handle an insect unless you know which type of beetle they are. Your chances of being sprayed are very slim, but it's never a bad idea to be cautious but not scared. It's a delicate balance.

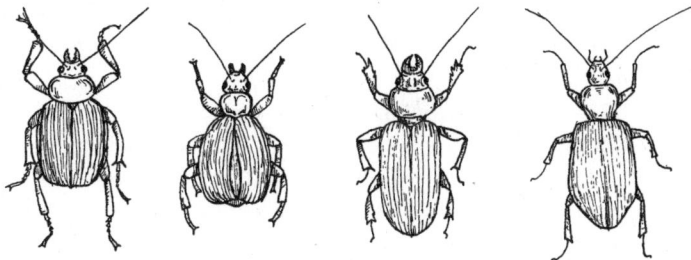

The left three beetles are different types of common ground beetles, which often come in black or brown. The one on the right is a bombardier beetle, which normally has an orange body with blue-green shimmery elytra.

Tiger beetles are an interesting species that belongs to Adephaga. They are very unique in terms of their look compared to other beetles.

Family Cicindelidae, genus *Lophyra*. This one can be found in Tanzania.

Their legs are delicate and skinny unlike the typical beetle leg, almost wispy. They have very large eyes and wicked looking mandibles (another word for mouthparts). Some other beetles, like wood borer beetles and stag beetles, have large mandibles as well; however, borers don't use them for eating and stag beetles use them for mating rituals. Tiger beetles use their mandible to expertly catch prey. Adults and larvae are carnivorous.

Tiger beetles can also move very quickly and can cover upwards of 53 times their body length per second. So imagine you're s6 feet tall. If you could cover the same amount of ground, you could run 318 feet in 1 second. That's pretty fast. Scientists at Cornell University discovered that these beetles run so quickly when hunting prey that they go blind momentarily. Entomologists noticed that tiger beetles would run in quick bursts, stop-and-go style. They determined they did this because they regain their sight when stopping, to check that their prey is still there.[44] The same entomologist at Cornell, Dr. Cole Gilbert, found that they use their antennae at high speeds to determine obstacles in their way and are able to respond in a flash to sticks or anything that is in their path.[45]

The adults are impressive hunters, and so are the larvae. They dig deep burrows and then wait at the entrance for prey to come along. They have large mandibles and hook themselves into the burrow to maintain stability. When something walks by, they attack and drag the prey back into the burrow with them. In order to build these burrows, the larvae need sandy areas to easily dig into. Adults also live in sandy, open habitats.

44 Gilbert, C. (1997). Visual control of cursorial prey pursuit by tiger beetles (Cicindelidae). *Journal of Comparative Physiology A*, 181, 217-230.
45 Zurek, D. B., & Gilbert, C. (2014). Static antennae act as locomotory guides that compensate for visual motion blur in a diurnal, keen-eyed predator. Proceedings of the Royal Society B: Biological Sciences, 281(1779), 20133072.

Most of the beetles I've talked about thus far are terrestrial, meaning they live exclusively on land. There are quite a few species of beetles that are aquatic or partially aquatic. In the introduction to this book, I mentioned the Cherokee origin story about the universe: the diving beetle from this suborder was the one they were referencing. Diving beetles are found in bodies of freshwater and have been given the title "best insect swimmers." They have powerful, flat, paddle-like hind legs and bristles on each leg to propel them forward or make sharp turns at a moment's notice. The adults and larvae live in the water for their entire lives. Even though they live underwater, they need to breathe oxygen. Adults do this by holding a bubble of air under their wings and breathing it in through their spiracles (which are little holes in their abdomen, since they don't have lungs). Adults can fly if they need to and that is how they disperse to other habitats that are suitable for them. All stages of these beetles are predaceous.

Many beetles in this order are aquatic, including the adorable (in both name and appearance) whirligig beetle. I remember watching them in the creek by my house when I was a kid. They swim like they're drunk and don't seem to worry much about hiding. Similar to water beetles, they have an oval body. They also are able to fly, eat other insects, and carry a pocket of air with them under their wings. Their leg movements, however, are dissimilar, which is why the water beetle swims so smoothly and the whirligig beetle looks like it just realized it had legs. They have evolved to have eyes that are split for viewing above and below the water at the same time.

They are also often easy to see when swimming on the surface because they swarm together. When they do, it reminds me of the nano-bot technology you see Iron Man using in the Avengers movies. They all seem to predict the movement of the other beetles around them, and move fluidly, in unison. Scientists have discovered that these creatures have an internal radar to detect what is around them to help make sure they don't run into each other.

How Humans Can Interact with Tiger, Ground, and Diving Beetles

Some of the coolest types of these bugs, like many tiger beetles, are currently threatened or endangered. The sandy areas they call home are being transformed into residential or commercial areas, sometimes being paved for roads. Off-roading causes issues when driving over the burrows of the larvae. These actions lead to less habitat for the adults and higher likelihood of death for the larvae. The Xerces Society is trying to create laws and conservation plans to help save the species that are most affected. They have found that in areas where other protections are in place for a species, like a threatened bird or mammal, it has positive effects on the bugs in that area as well.[46]

But it's not only the terrestrial insects that are in trouble. Some aquatic species, like the water beetles, have dwindling numbers. Their populations are threatened because of the reduction in freshwater sites and fragmentation. Even though they are able to fly, if suitable locations are farther and farther away, they are unable to spread. They also do not do well in freshwater locations that contain fish. Fish are major predators and greatly reduce the number of beetles that survive. Since these bugs live in water, they can be useful as bioindicators in research to determine the toxins and quality of an area. Some places breed water bugs for eating, although the practice is falling out of use since they require a lot of care and yield little sustenance. Some people don't have interest in eating them but use them for other purposes. According to entomologists Kelly Miller and Johannes Bergsten, "in areas of eastern Africa, young girls collect diving beetles that are induced to bite the nipples, which is thought to stimulate breast growth."[47]

How Human Culture Is Shaped by Ground, Tiger, and Diving Beetles

During the Victorian era, when people loved putting fireflies in their hair and beetles on their collars, they also loved collecting bugs for large, personal insect collections. Beetles have always been fascinating to collectors simply because there are so many different species and many of them are morphologically distinct. Apparently, Charles Darwin had a fondness for beetles. In his autobiography, he talks about how he would blow off what he was supposed to do to go chase ground beetles.[48] People still maintain personal insect collections, and based

46   Xerces Society. (n.d.). Tiger beetles. xerces.org/endangered-species/endangered-beetles/tiger-beetles

47   Miller, K. B., & Bergsten, J. (2016). *Diving beetles of the world: Systematics and biology of the Dytiscidae.* JHU Press.

48   Darwin, C. (2010). T*he Autobiography of Charles Darwin.* (Reprint). Prometheus Books.

on the sheer number of beetles, they are present in every collection. Almost every insect I pin while volunteering at the science museum is a beetle.

Beetles are truly wondrous creatures who, even while having the necessary similarities to be in the same order, are so diverse in form, habitat, diet, and behaviors. This information only barely touches the deep, complex world of beetles. There are numerous books that cover just this topic so there is no shortage of material for those who want to learn more about them. Along the way, keep in mind that beetles are rarely pests and integral to the world around us. Next time you see one inside, ask yourself if smashing it is necessary. Maybe just put it outside and create spaces out there that support their success.

Quick Facts: Ground/Tiger/Diving Beetles

- Can eat as much food per day as they weigh.
- Many beetles are aquatic, like whirligig beetles.
- Aquatic beetle numbers are decreasing because of lack of freshwater sites and fragmentation.
- Terrestrial beetles (those that live on land) have decreasing numbers because of fragmentation and habitat destruction.
- All of these beetles have hard outer wings (called elytra) and soft underwings.

Bug Spotlight: Tiger Beetles

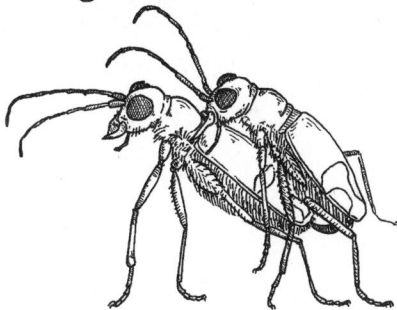

During mating, the male tiger beetle uses his mandibles to hold onto the female. Their massive eyes and mandibles help them when hunting.

- One of the best beetle hunters.
- They run so quickly that their eyes can't process light and they lose the ability to see while running.

- The larva are vicious hunters who attack their prey from underground.
- Some species mimic the sound of toxic moths to dissuade bats from attacking them.
- They live in areas with loose dirt and are threatened by habitat destruction (from building apartments to tires driving over the dirt which then trap the larva).

QUICK SHOUTOUT: The Other Suborders

Archostemata: This is the smallest suborder of beetles, only having forty-five living species. These beetles are closely related to many extinct species and are considered rare.

Myxophaga: This suborder has about sixty-five species. All species are aquatic or semi-aquatic. They feed on algae. They appear on every continent apart from Australia to Antarctica.

Protocoleoptera: All of these species are extinct but preserved in fossil records.

Part IV The Most Misunderstood: Bees, Wasps, and Ants (Order Hymenoptera)

*T*recently attended a pollinator summit where one speaker talked about the importance of wasps and why they are so cool. Even in a room full of bug lovers, there seemed to be some wariness in accepting wasps into their hearts. Wasps and bees are part of the same family, but one is appreciated for its utilitarianism and fuzzy butts, the other is feared and disliked. Ants, who are also related, fall somewhere in the middle at most times. Species in this family run the gamut in terms of public opinion. Opinions aside, they are all important economically and ecologically.

The name *Hymenoptera* means "membrane wing" (*humen* + *pteron*). Looking at their wings, you can see that unlike butterflies and beetles, hymenopterans have clear wings with noticeable veins. They do have two sets of wings, but the top and bottom wings are kind of velcroed together, and when they move or land, the wings go up and down in unison. This phenomenon is not common among other insects, so it is unique to this order. Their wings beat so quickly that when they're flying, it's not easy to notice the wings being together, but slow-motion cameras make this reality apparent.[49]

Over the last few decades, people have started to gain interest in the wellbeing of pollinators like bees. Scholars noticed a shift in attention when stories of colony collapse were published in popular newspapers and magazines. Since then, the public's view of bees has changed: we love their cute, fuzzy butts; we appreciate their role as pollinators; we put up more bee houses and grow more pollinator gardens. However, wasps and ants were left out of the ecological pollinator campaign, so the love for them is still lukewarm at best. They do still benefit from the creation of pollinator-friendly spaces. Wasps are pollinators. They may not be as effective as bees, but some plants rely exclusively on wasps to pollinate (more on that later). Ants are not pollinators but

---

49 NOTE: If you ever want to entertain yourself, look up Dr. Adrian Smith's slow-motion recordings of insects taking off. So cool.

are extremely important ecological forces and can frequently be found on plants.

Apart from their ecological importance, hymenopterans are important in culture as well. The mascot situation would be sad without wasps, as would the selection of Pokemon. And without the bugs in this order, we'd miss out on amazing comic book characters like Ant-Man (Marvel Comics), the Wasp (Marvel Comics), and Bumblebee (DC Comics). Three hymenopterans were in the same movie and comics together: Ant-Man, the Wasp, and Yellowjacket. Two of them are heroes (Ant-Man, the Wasp), and one is a villain (Yellowjacket). In the 2015 Ant-Man movie, the accuracy of the species of ants and their behaviors was impressive. The movies were better than the comics at showcasing species. They also made the audience feel empathy towards ants in a way that many films don't. Six thumbs up for *Ant-Man* (2015).

In music, wasps, bees, and ants show up more than any other insects. Joseph Coelho, a biology professor at Quincy University, combed through 1,338 songs that mention insects. Of those songs, 25.4 percent were about hymenopterans. He attributed them being the most commonly cited "because they have both charming qualities and painful associations, such as stinging. Not surprisingly, these applications are similar to those found among films."[50]

Even though bees may be the most loved of the hymenopterans by the general public, understanding the place of wasps and ants in the ecosystem and culture should help create some appreciation around them. Especially since most of them (like other orders of insects) are not human pests.

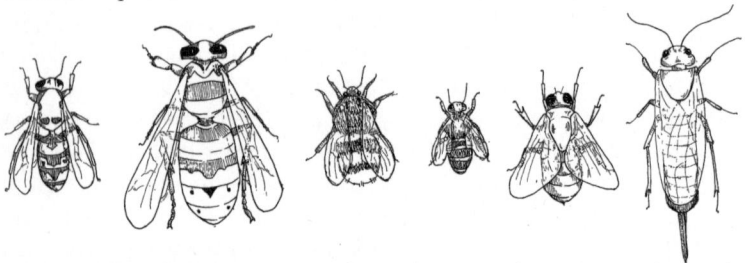

From the left: Yellowjacket (genus *Vespula*, wasp), European hornet (*Vespa crabro*, wasp), bumblebee (genus *Bombus*), European honeybee (*Apis mellifera*), common house fly (*Musca domestica*), giant horntail with noticeable ovipositor (*Urocerus gigas*, wasp)

---

50 Coelho, J. (2000). Insects in rock & roll music. *American Entomologist*, 46(3), 186-200.

CHAPTER 9: BEES (CLADE ANTHOPHILA)

*W*hen people think of bees, they often think of honeybees. *Apis mellifera*, the Western (or European) honeybee, showed up in fossil records about two million years ago, which is only about a tenth as long as other *Apis* species who live in open nests (and only survive in the tropics). European honeybees are not native to the U.S., but since being introduced here by European farmers, make up a large part of agricultural pollinators. They are the species most often used in beekeeping. People have been keeping bees for upwards of 9,000 years, although the practice has shifted over time. Originally, the entire hive would die after a single season. The hives were often kept in jars or clay pots, and all of the honey was removed from the nest, which would kill the bees. It wasn't until the seventeenth century that bees could be kept without killing the entire hive. In the 1900s, advancements were made that ensured that bees would live for many years.[51]

And since that time, bees have come so far in the public eye. According to many scientists I have spoken to, the public became particularly interested in bees with increased reports of colony collapse. Colony collapse occurs in honeybee hives, causing significant and often fatal problems. Different mites or diseases can be the underlying issues, which weaken the hive and make the bees more susceptible to serious issues. This travesty created more appreciation around pollinators, especially when people began to realize how important they are to our survival with regard to food and to our world in general.

How Bees Shape Our World

Pollinators help in the production of nearly 75 percent of all crops and roughly 80 percent of all flowering plants. According to the White House, pollinators contribute more than 24 billion dollars to the United States economy. In 2009, the crop benefits from native insect pollination (i.e., NOT honeybees, more later) in the United States were valued at more than 9 billion dollars. Most fruits, nuts, berries, and

---

51 Buchmann, S. L. (2023). *What a bee knows: Exploring the thoughts, memories, and personalities of bees*. United States: Island Press.

other fresh produce require insect pollinators. Foods like chocolate, vanilla, coffee, almonds, and berries also wouldn't be available without insect pollinators. Even dairy products would be harder to come by, as many cows eat diets of pollinator-dependent alfalfa or clover.

Even crops that self-fertilize benefit from insect pollination. Most strawberry cultivars produce tiny flower clusters including both male and female, and sometimes sterile, flowers. While wind will sometimes allow these flowers to share pollen, open pollination by insects will increase fruit set, yield, and quality. As many as twenty bee visits to each flower cluster are required to fully pollinate all of the strawberry flowers. Without pollinators, our diet would quickly be limited to a few wind-pollinated crops such as corn and wheat. A nature walk or stroll through a garden would be a very different experience without pollinators.

For hunters, fishing enthusiasts, and bird-watchers, the benefits of insect pollinators are clear. Fruits and seeds derived from insect pollination are a major part of the diet of mammals ranging from red-backed voles to grizzly bears. More than 90 percent of birds rely on insects at some stage in their life. Studies have shown that diverse and abundant plant communities that support pollinators also support the types of insects that are favored by fish.

In addition to all of these products, pollinator services are a multi-billion dollar industry. Honeybees alone are linked to $14 billion worth of services in the U.S. alone. When they go to a flower, they collect and store pollen on their legs to take back to the hive. Honeybees and other bees have different ways of storing that pollen. Some have what is called a "pollen basket" or corbicula. Others, like some mason bees, have what is called a scopa; they don't have a basket but instead have long hairs on their legs or abdomens to hold pollen. Interestingly, honeybees are one of the few species to have hives. They need to bring food back for a large number of larvae. Bumblebees may live in small groups but the hive is much different looking and is often underground. 90 percent of bees live either in the ground or pithy stems (the old stems of those sunflowers in your neighborhood or those bee hotels). The bees who don't live in hives need only to feed themselves and their young.

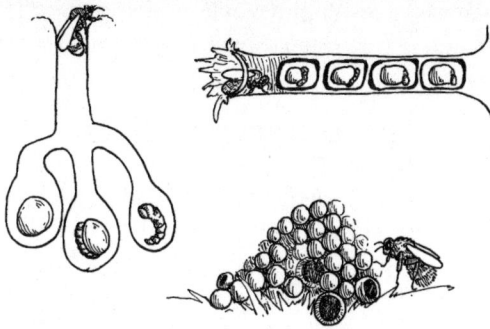

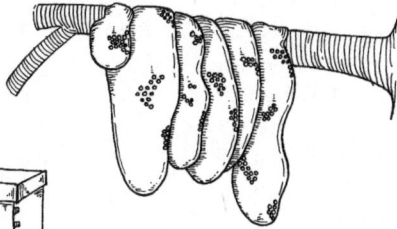

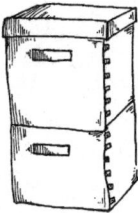

Different types of housing for bees. Top left: solitary ground dwelling bee. Top right: stem/wood nesting bee. Middle: an example of bumblebee "hives," which are often out in the open and are shared by a few bees. Bottom left: what the combs look like in a hive; what a hive looks like in the wild; and the hive boxes that are used by beekeepers (apiarists)

Hundreds of native species are great pollinators, like the blue orchard bee. Bumblebees use "buzz pollination," which is when they shiver their flight muscles to create strong vibrations that shake pollen loose. Tomato plants have sticky pollen that needs to be shaken, so bumblebees are the most efficient pollinator of these plants. All bees and many other invertebrates are negatively affected by pesticides. Xerces Society warns against the use of pesticides with neonicotinoids. These pesticides were once considered better since they were less toxic to mammals and other vertebrate animals; however, it has been discovered that neonics are highly toxic to pollinators. If you want to plant a pollinator garden, look for seeds that aren't treated with neonicotinoids.[52]

How Humans Can Interact with Bees

Many people believe that keeping bees is good for conservation efforts. Unfortunately, that is not the case. The rise in hobby beekeeping, which dramatically increased during COVID lockdowns, followed strong campaigns to "save the bees." According to Xerces Society, there are currently over 3 million hives in the U.S. alone. But honeybees are least in need of saving. They are probably doing more harm than good. When I first learned this, it sent me reeling. How could that be possible? I felt like my entire life I had been told that honeybees were the most important bees in all the land.

I was sorely mistaken.[53] There are 16,000 known bee species worldwide. North America is home to 3,600 species of bees. Over 90 percent of species are solitary and make homes in dirt or pithy stems. There are several issues between introduced species (more specifically *Apis mellifera*) and native species of bees. Honeybees are connected to decline in native bee populations, both because they fight for food and pass on viral infections to native bees when visiting the same flowers.[54] Honeybees have been found to spread Deformed Wing Virus (DWV) to wild bumblebees, as well as passing on a fungal infection. DWV is spread by mites that are associated with colony collapse. DWV seems to have been passed to native bee populations who cannot handle the disease as well. The fungal infection (called Nosema) is also less devastating to honeybees compared to bumblebees.

Wild bees are vital for crop pollination and increase yields across many crops. They often are more efficient at pollinating crops native to North America than honeybees. For example, a honeybee would have to visit a blueberry flower four times to deposit the same amount of pollen as a single visit from a bumble bee queen. Entomologists and conservationists are encouraging agriculture to rely more heavily

52 Xerces Society. (n.d.). *Understanding Neonicotinoids*. xerces.org/pesticides/understanding-neonicotinoids

53 Mcafee, A. (2020, Nov 4). The problem with honey bees. *Scientific American*. scientificamerican.com/article/the-problem-with-honey-bees/

54 Alger, S. A., Burnham, P. A., Boncristiani, H. F., & Brody, A. K. (2019). RNA virus spillover from managed honeybees (Apis mellifera) to wild bumblebees (Bombus spp.). *PloS one*, 14(6), e0217822.

on native populations of pollinators. Still, honeybees get all the love: seventeen states have the Western honeybee as their state insect. I'm going to guess that the percentage of people who think honeybees are native is extremely high. Like I said, I was in that category until fairly recently.

Even though honeybees may not be the best pollinators compared to their native counterparts, they are extremely important to humans, especially economically. The number of bee products made and sold are extensive:

- Bee venom: Commercially available and used to create tolerance in people who are allergic. Some people also use the venom for arthritis. The process is often used with a live bee who then dies in the process.
- Bee brood (specifically *Apis mellifera*): Immature bees (also called larvae) are regarded as a delicious snack in some parts of Asia.
- Beeswax (specifically *Apis mellifera*): Used for candles and in models for bronze statues and gold ornaments, ointments, emollient skin creams and lotions, polishes and protective coatings, armament lubrications, and electrical engineering. In 2022, more than 65,000 tons of beeswax were produced worldwide.[55]
- Pollen: Used as a dietary supplement to humans and domestic animals. It is used by bees to boost brood production. Pollen is used in plant breeding programs and the study/treatment of allergic conditions like hay fever. Compared to some other bee materials, pollen is easy to collect by using a pollen trap at the entrance to the hive.
- Propolis: A resin-like material used by bees to make adjustments or fix cracks in hives. Humans use it for cosmetic and healing creams because it is believed to fight against infection and have anti-inflammatory properties. It can also be used in chewing gum.
- Royal Jelly (specifically *Apis mellifera*): Used by bees to rear their queen. Humans use it for combating bacteria, for prolonging life, and as an aphrodisiac. None of these uses

55 Food and Agricultural Organization of the United Nations. (2022). Beeswax Data, Worldwide. [Data Set] fao.org/faostat/en/#data

have been substantiated in research. Extraction of royal jelly from the hive is extremely labor intensive and time sensitive; therefore, it is often expensive and many products that say they are royal jelly may be fake.

- Mead: An alcohol created using honey. The name is a combo of the words *medicine* and *madness*. Mead is a sweet liquor with a somewhat viscous texture and has a higher alcohol content, similar to wine. It is also believed to have been a popular alcoholic drink before wine was created.

How Human Culture Is Shaped by Bees

A horticulturist I spoke to in Denver said that people often ask how they can plant a pollinator garden that doesn't attract bees. The answer is that you can't have a bee-free garden. The fear of these effective pollinators is based on a lack of understanding. People don't often like it when bees fly around them because they are concerned about getting stung. Good news is that most species have stingers too weak to penetrate human skin, and male bees don't have stingers. Of course, it's hard to tell which one is male and female when observing them midair.

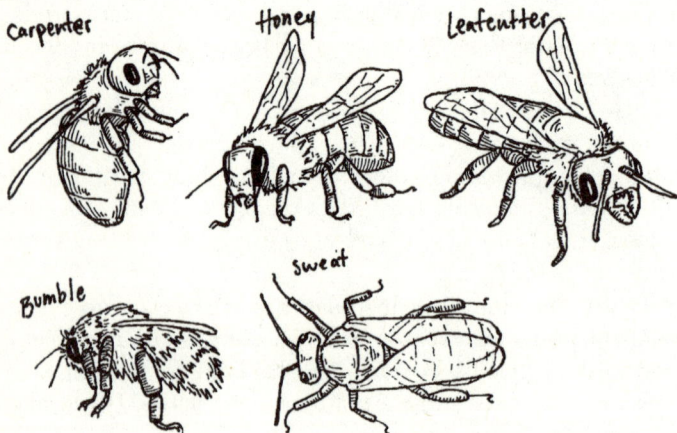

Bumble bees are very fuzzy whereas carpenter bees do not have furry abdomens. Leafcutter bees have large mandibles, and sweat bees are often tiny and may be brightly colored.

Culturally, bees hold some very important distinctions. In the Quran, the honeybee (perhaps a native Asian variety) is the only creature to talk directly to God.

And your Lord revealed to the bee saying: Make hives in the mountains and in the trees and in what they build: Then eat of all the fruits Lord submissively. There comes forth from within it a beverage of many colours in which there is healing for men (68-69).[56]

In the Dead Sea Scrolls, bees were forbidden to be eaten because bees were considered unclean because they were believed to be born out of dead bodies of animals. Some flies are bee mimics, so this idea may have come from a misidentification. However, the vulture bee does eat meat and has a blood red hive. I recommend you look it up. It's pretty crazy.

In literature and music, bees have captured the attention of creatives. The term "busy as a bee" was first used in Chaucer's Canterbury Tales from 1392ish: "In wommen been! For ay as bisy as bees." Think of all the other phrases referencing bees: beehive hairdo, bee's knees, bee in her bonnet, the list goes on. One of the most famous and recognizable pieces of classical music is called "Flight of the Bumblebee" (1900) ,which tried to capture the chaotic buzzing of a bee. Most people can recognize the tune.

More recently, Cole Porter sang about "the birds and the bee.s" The song is just about how all animals, regardless of size, have sex. Unfortunately, that was only a rebrand of the song to remove some amazingly racist language. (Thanks, CBS, for suggesting the changes.)

In animation, bees are a few characters in the original Pokemon line-up. Kakuna and their evolved form, Beedrill, are the pupae and adult forms of bees. I'm sure that most people can think of multiple examples of bees in culture that I have not mentioned. There are so many.

Quick Facts: Bees

- The majority of bees are solitary, meaning they don't have hives.
- Most bees live in the ground or in stems.
- Honeybees are not native to the U.S.A. The ones that people use in beekeeping are from Europe.

---

56 Ali, M. (1920). The Holy Qur-án: Containing Arabic Text with English Translation and Commentary. India: Ahmadiyya anjuman-i-isháat-i-Islam.

- Native bees are much better pollinators than honeybees but can sometimes be outcompeted for food sources because of the number of honeybees present in the us.
- Beekeeping is NOT a conservation effort. If you want to help bees, plant native neonicotinoid free flowers that attract bees and have space for them to hang out in your yard.

Bug Spotlight: Bumblebees (genus *Bombus*)

Bumblebees will store most of the pollen they collect on their corbiculae located on their legs, but because of the buzz pollination and general fuzziness, lots of additional pollen gets stuck to their bodies.

- Bumblebees are eusocial, meaning they live with a couple others in a hive.
- Their hive looks very different from a honeybee hive and does not contain nearly as many pupas.
- They use "buzz pollination" to remove sticky pollen from flowers, like in tomato plants, making them more effective pollinators. They do that by shaking their flight muscles fast and hard to dislodge hard-to-get pollen.
- They are not likely to sting you and are overall pretty chill. If they do sting, which should only happen if you step on them or squeeze one in your hand, they do not lose their stinger.
- There are thousands of species of bumblebee and a few in the U.S., like the American bumblebee, are moving closer towards endangered status. This is because of pesticide use, habitat loss, and climate change.

CHAPTER 10: WASPS (SUBORDER APOCRITA)

Wasps are dope, but they often get a bad rap. Understandable considering the stinging aspect; however, they are misunderstood much more than bees. Bees, at least in most of their forms, are appreciated for utilitarian and environmental purposes. Wasps? Not so much. But they do serve important functions in the ecosystem. Wasps are the reason we have ink and paper: galls created by wasps on oak trees are used to make iron gall ink. Paper wasps inspired Rene de Reaumer to use wood for papermaking. They are also exceptionally talented in pest management. Parasitic wasps keep fly, caterpillar, and other insect numbers in check by parasitizing them. There are even parasitic wasps of parasitic wasps, called hyper-parasites. A particular species of moth is parasitized by a wasp, and that wasp is then parasitized by a different species of wasp.

How Wasps Shape Our World

Waps are so influential in our ecosystem: most insects have a parasitic wasp that targets them specifically. Braconid wasps are parasitic to caterpillars, especially hornworms. They lay their eggs in the caterpillar, resulting in very visible wasp cocoons on the parasitized bug's body.

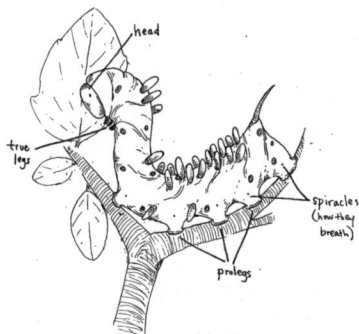

The small egg-like structures on the back of the tomato hornworm caterpillar (*Manduca quinquemaculata*) are eggs laid by parasitic wasps. The larvae will hatch and eat the hornworm while it's still alive.

These parasitic relationships are important and when very specific, linked to concerns of coextinction. This is due to the fact that these organisms may depend on a single species throughout their lives in comparison to multiple hosts. If that single host disappears, so might its parasites. It isn't a guarantee, as some wasps may be more evolutionarily flexible. Many scholars in sustainable agricultural practices encourage the use of parasitic wasps as they can be very effective and cheap and also can reduce the amount of insecticides used. Introducing new species is often discouraged as scientists are unable to fully predict how their presence will disrupt the native ecosystem. Even non-parasitoid wasp introduction can be a problem. In New Zealand, yellow jacket wasps were introduced and are now in a food conflict with a native parrot species, leading to declining numbers in the parrot population.

When they are in their native habitats, wasps' ecological contribution can be integral to the agricultural stability of some plants. Figs would not be able to grow without the fig wasp. They co-evolved to need one another: figs are only able to grow because of their relationship with the wasp. Most figs have one or two species that fertilize that particular species of fig (750 species of fig worldwide). Female wasps pollinate female fig flowers, as well as lay eggs there. Some weeks later, the wasp offspring hatch just as the male fig flowers have matured their pollen sacs. The new generation of female wasps leave the plant through holes made by the males and carry pollen to plants elsewhere.[57]

Another example of a specific relationship is between hammer orchids and wasps. These orchids look like and smell like female wasps. The male tries to mate with the flower, thinking its a female wasp, and when they attach, they get orchid pollen on them. When they try it again with another orchid, they pollinate that orchid.[58] The videos of the process are fascinating. I feel a little sorry for the wasps because they appear unable to tell the difference. It is an effective, if not precarious, relationship. Those orchids rely heavily on those wasps. Without them, pollination would be close to impossible.

57 Cook, J. M., & West, S. A. (2005). Figs and fig wasps. Current Biology, 15(24), R978–R980.
58 Menz, M. H., Phillips, R. D., Dixon, K. W., Peakall, R., & Didham, R. K. (2013). Mate-searching behaviour of common and rare wasps and the implications for pollen movement of the sexually deceptive orchids they pollinate. PLoS One, 8(3), e59111.

How Humans Can Interact with Wasps

Wasps have a really scary reputation, leaving us humans unsure how to interact with these bugs. Similar to bees, many wasps are solitary. They may burrow in dirt or sand and have a single female per burrow. Cicada killers are large solitary wasps that dig in dirt and drag cicadas back as food for their young. They can be quite intimidating but are unlikely to sting (they really want to avoid you at all costs). Many parasitic wasps are solitary, laying their eggs in a burrow or a live insect, who then cares for the eggs. Most wasps live this way. Some are somewhat social (like paper wasps) and even fewer are eusocial (meaning similar behavior to honeybees). One of those species is yellowjackets. They can create fairly large hives in trees, attics, or other human-built spaces. Unlike honeybees, wasps do not lose their stinger or die after stinging, meaning they can sting multiple times. They do defend their nests fairly aggressively, so if you need a nest removed because it is too close to where you live, contact a professional. Using insecticides is discouraged since it can cause further issues for you and the environment.

Wasps most commonly encountered by people in urban areas are seen as pests and can be territorial. When you hear the term "as mad as hornets," a vivid memory may appear of that one time your dad ran over the hornets' nest with a lawnmower. In general, most wasps, including common urban species like paper wasps, do not want to hurt you, even if they fly around your head. If at all possible, let them be. They are important for the ecosystem and don't want any confrontation.

How Human Culture Is Shaped by Wasps

Galls are an abnormal growth that a plant produces when the plant senses an insect invader. They can be created in reaction to many types of insects. Gall wasps, however, are the insects responsible for causing plants to react and create oak galls. These oak galls were used in creating ink for many centuries. *Beowulf* and the Magna Carta were both written using gall ink. The wasp lays their egg in the tree, and the tree reacts by creating a gall around the wasp to avoid damage to the tree. The wasp then hatches out of that gall and the process continues.[59] There is a species that parasitizes gall wasps and can also

59   Knight, B. (2013, Aug 19). *Iron gall ink and wasps*. British Library. blogs.bl.uk/collectioncare/2013/08/iron-gall-ink-and-wasps.html

be found in galls. The ink itself is susceptible to "ink corrosion" so collections that include documents with oak gall ink (which could be anything written from the Middle Ages to the twentieth century) need to be vigilant about keeping those documents from molding.

We have the ink, so now we need the paper, which is also possible because of wasps. We had paper but the process was always shifting as people tried new ways of making it. A French scientist, Rene Reamur, stated how impressive wasp nests were at sticking around forever. This idea wasn't immediately accepted but during the early 1800s, humans tried to make paper using tree pulp. Initially, it wasn't the best but over time, has improved. We still can't do it nearly as well (or as easily) as paper wasps do.[60]

Finally, waps in general are the inspiration for some pretty cool characters. Janet van Dyne, also known as the Wasp in Marvel Comics, is a very positive representation of these often-unloved bugs. She is able to shrink down to wasp size and she has little electric shocks that are like stings. Or take Yellowjacket, a villain in the Marvel comics that often comes up against Ant-Man and the Wasp. His abilities include two over-the-shoulder stingers, as well as flying and shrinking.

Quick Facts: Wasps

- Like bees, the majority of wasps are solitary and many of them are parasitic.
- Parasitic wasps lay their eggs with or inside of live/paralyzed prey.
- They help keep down numbers of pest species.
- Without wasps, we wouldn't have ink and probably wouldn't have paper (in the same way at least).
- Many wasps are incapable of stinging and have an ovipositor (for pumping out eggs) instead.

---

60 Leinhard, J. H. (n.d.). *Of wasps making paper*. University of Houston. engines.egr.uh.edu/episode/1052

Bug Spotlight: Velvet Ant

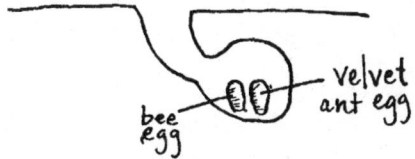

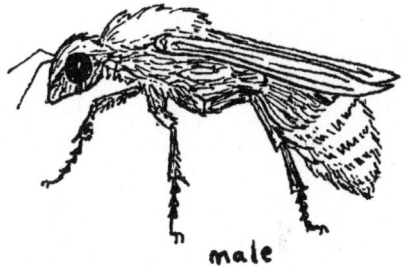

female

bee egg

velvet ant egg

male

Family Mutillidae

- Looks like an ant, definitely isn't.
- It is fuzzy so you may want to pick it up. DON'T. Their sting is extremely painful.
- They lay their eggs in the same chamber as a ground dwelling bee's egg. The velvet ant larvae hatches first and then eats the bee for sustenance.
- Males look more like a wasp and have wings.
- Over 7,000 species worldwide.

Chapter 11: Ants (Superfamily Formicidae)

*A*nts are at an interesting intersection between tolerated and hated. Most of the time, ants are seen as unimportant, like when you pass them going about their business on the sidewalk: noticed but unthreatening. In certain situations, people despise ants: that time of year when thousands fly to find new colonies, when they won't seem to leave your house, when they are the biting kind that hurt like hell. Basically, they are most often noticed when they are pests.

And it's understandable from a human perspective. When ants fly away to find their own colonies, the numbers can seem overwhelming. Most of them will be food for animals; only some will be successful in starting colonies. Along the way, they are often mistaken for termites, but unlike termites, they pose no threat to humans or buildings. The ones that bite (and it hurts like hell) are not native to the U.S.A. Most native species can bite but don't cause any lasting issues, more like a little pinch. Fire ants however are a whole different story.

How Ants Shape Our World

Fire ants were introduced to the U.S., most likely by accident, but have now set up all across the south. The non-native species has been extremely successful and has even caused some serious issues for native species, outcompeting them for food and space. The United States government tried to eradicate them. The campaign cost $200 million and didn't do anything they wanted it to do. It actually killed large numbers of beneficial bugs, including native species. Fire ants are here and there isn't anything we can do about it. Some states even have Fire Ant Festivals, as a way to acknowledge the difficult co-existence that won't be changing any time soon. Anyone who has been bitten by one (or the dozens who swarm) will tell you that it is terrible. Most often they are disturbed when the nest is stepped on. They move quickly and can sting multiple times.

Another invasive ant has found itself on Christmas Island. That island is known for its red crabs that make large migrations across the island each year. Unfortunately, the crazy ant, as it's called, is devastating red crab populations. They will attack small and baby crabs to kill and eat. With dwindling numbers of crabs, the people who live there are noticing a difference in seedling recruitment and litter breakdown. Those tasks are normally completed by crabs, and with less of them, they are seeing more litter throughout the island. It is having a ripple effect throughout the Christmas Island ecosystem.[61] They may be introducing parasitic wasps to kill the crazy ant population.

Despite all the trouble they cause, the above invasive species are only a tiny fraction of the number of ant species, which is estimated at about 22,000. While there are more than these two species that are considered annoying, even with those added, the annoying count only amounts to less than 1 percent of species. Most of the ants you see in your house are not there to set up shop. They do enjoy sweet foods, like honey, so be sure to keep your counters clear of easy to access sweetness. (I remember as a kid having a honey pot on the counter and finding a trail of ants leading to and away from it.) They are opportunistic omnivores and are known to eat other insects, dead animals, sugar, fungus, and seeds.

Similar to other bugs, the majority of ants are integral members of the ecosystem. Ants can lift ten times their body weight. For me, that would be like me lifting 1600 lbs. (I have trouble lifting 50 lbs.) Some scientists believe that certain species can lift almost 50 times their body weight. They are super strong regardless. I've seen them work together to move a cicada, which was about five times their size, to their dwelling. Most of the ants that humans see are worker ants which make up the largest percentage of the colony and are female. Colonies also have different types of members: one queen, multiple sterile females who are workers and soldiers, and sporadic males for mating. Some of the soldier ants in different colonies are noticeable because of their abnormally large head compared to the others in the colony. Queens are also large compared to the rest of the ants.

61   Bittel, J. (2015, Dec 24). The Christmas Crab Massacre. NRDC. nrdc.org/stories/christmas-crab-massacre

Leafcutter ants belong to three different genera: *Atta*, *Acromyrmex*, and *Amoimyrmex* that live all across the Americas.

Their nests must be large enough for the entire colony and can be extensive. Some colonies can have over hundreds of millions of ants while some may have only five hundred or so. There are also supercolonies, which is when two nests meet and the ants are not competitive with each other. Some ants will also take over another colony, essentially creating slaves to do their bidding. They replace the queen and use the ants from the previous colony to do all of the work. The nest has multiple chambers which each serve different functions. Some are for broods (also known as babies), some are for food storage, and some are for defense. Food storage can look different for different colonies. Some store seeds, some tend fungus in the nest chambers, and some store honey. Honeypot ants store honey in the abdomens (or butt) of the worker ants to save for later. The ants of one colony all smell the same, which serves as an indication that they belong in that nest. Ants with unfamiliar smells are often attacked.

These small creatures beneath the surface have a huge impact on the soil, decomposition, and thriving world under our feet. We may see them, but we don't often see how they shape our world.

How Humans Can Interact with Ants

The bullet ant, who is native to South America, has the most painful sting of any insect. It feels like getting shot, hence the name.[62] They do not seek out humans and only sting when stepped on or feeling attacked.

Some ants are used in medicine, like carpenter ants. They have large heads and mandibles, so they have been known to be used as stitches in the Mediterranean, northern Africa, and India. Once their jaws close, it is difficult to open them, which keeps the wound closed.

Ants are human-like in a surprising way: They are known to farm animals to their benefit. Aphids, which are often considered

62 NOTE: The tarantula hawk, which has the second most painful sting in the world, does live in the U.S. but is super chill and rarely stings humans.

agricultural pests, produce a sugar-like substance that is desirable to ants. Ants farm the aphids to collect this sugar, protecting the aphids from predators like ladybug larvae.

Ants have been to space where scientists tested their navigation abilities. They can find their way home from extreme distances even in zero gravity. Scientists are trying to determine how these capabilities could be used for our benefit. If you see an ant wandering by themselves, know they will most likely be able to find their way back to their home (and won't set up shop in yours).

They even have devastating wars. Pavement ants will have massive battles leaving hundreds dead on the sidewalk. In some colonies, even hundreds of deaths don't make a dent in their numbers. You may notice large clumps of ants on the ground, which is often an indication that they are at war. If you see them, just leave them be, as they are only attacking each other.

In Brazilian rainforests, the biomass of ants alone has been "estimated at approximately four times greater than the biomass of all the vertebrates combined."[63] They use their social network to be extremely successful hunters. They can inundate larger prey and drag them back to their nests. In those large numbers, they are very effective.

How Human Culture Is Shaped by Ants

Walter R. Tschinkel is an entomologist who pours aluminum casts of nests, which he then digs up to show the full structure. They look incredible and can be up to eleven feet deep. Unfortunately, the ants are killed in the process. Some artists try to use abandoned nests or nests of invasive species, like the fire ant. Many people disagree with the process. It does show the impressiveness of the nests and can be used as an educational tool. Not killing a colony sounds good too though.

The movie *A Bug's Life* (1998) does a great job of showing the hierarchy of a colony and the expansiveness of ant nests. Their social network is extremely advanced and can be different depending on the species. Similar to honeybees, the workers are mostly infertile females and a few fertile males who serve the queen; most babies are female. They have factory-like worker systems inside a fortress. Ant

63 Franks, N.R. (2009). Ants. In Resh, V. H., & Cardé, R. T. (eds.), Encyclopedia of Insects. (2nd edition, pp. 24–27). Elsevier.

queens, compared to other insects of similar size, live for extremely long periods of time, possibly over a decade (like termite queens). The female workers live for one to three years and the males only live for a few weeks.

Humans also use ants in creative ways. Surrealist artist Salvador Dali often used ants in his art. Humans also use them as literal and figurative representations in film. The first "big bug" film called *THEM!* was released in 1954. The plot included giant irradiated ants found in Arizona and represented the fear and uncertainty of nuclear war. In 1984, an entomology professor, Dr. May Berenbaum, at University of Illinois Urbana-Champaign started the Insect Fear Film Festival. The goal of these festivals is to attract audiences who watch the films, and then use those films to highlight popular misconceptions about insects with an expert panel. People can also handle live or pinned specimens between films. In 2023, it celebrated its fortieth season. In the *Encyclopedia of Insects*, Berenbaum states that "insect morphology in the movies reflects the relatively sketchy familiarity most filmmakers have with entomological reality."[64] In films, animators often change insects tto be more "attractive" to audiences: in *A Bug's Life*, the ants have four legs instead of six, to seem more human-like

Even before film, ants have long been a part of our stories. An old Cornish belief states that ants were indistinguishable from faeries. Faeries were souls of heathen people who were too bad for heaven but too good for hell, therefore they were trapped on earth forever. Their bodies changed to the size of ants. Because of this folklore, people thought that killing ants was bad luck. *Antsy* is a term dating back to the mid-1800s, but to "have ants in one's pants" didn't become a common phrase until the 1950s. It means to be fidgety and restless.

Humans aren't the only animals that use ants to their advantage: Some birds put ants on themselves to clean off parasites, called "anting." Acadia trees rely on ants as defense. The ants provide vital protection to the bull's-horn acacias day and night, and it has been shown that without the ants, *Acacia cornigera* suffer greater damage from attacking insects and tend to be overgrown by competing plant species. There are many other insects that mimic ants: grasshoppers,

64 Berenbaum, M. R. & Leskosky, R. L. Movies, insects in. In Resh, V. H., & Cardé, R. T. (eds.), Encyclopedia of Insects. (2nd edition, pp. 668-675). Elsevier.

spiders, flies, beetles, and even a plant has ant-like leaf patterns as a defense mechanism.

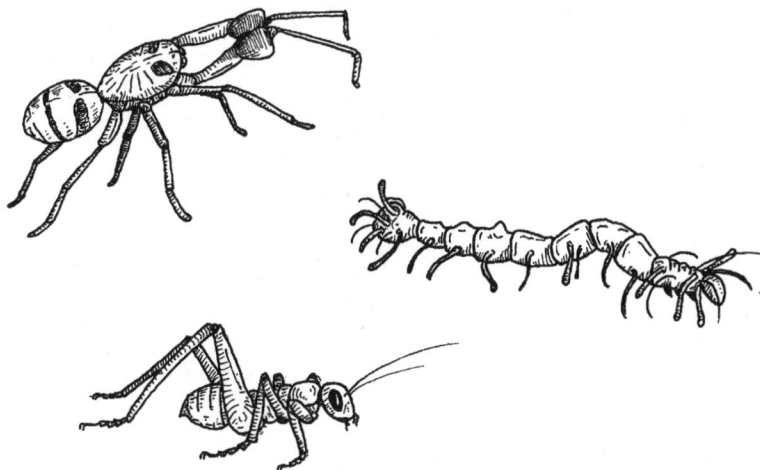

The spider mimic, *Myrmaplata plataleoides*, holds its front legs forward to form a head and antenna. The caterpillar, *Homodes bracteigutta*, looks like multiple ants following each other. The katydid mimic, genus *Macroxiphus*, has the color of ants to blend in, also having smaller back legs and longer heads than most katydids.

Quick Facts: Ants

- The biomass of ants is estimated at approximately four times greater than the biomass of all the vertebrates combined. Just ants.
- They were the first insects to appear in big bug horror movies.
- Ants have highly structured, complex social systems.
- Fire ants are not native to the U.S., but even after $200 million campaign to get rid of them, they are going strong.
- Ants have many similarities with humans: waging wars, farming (both aphids and fungus), and even performing surgery (the only other animal to be observed doing this).

Bug Spotlight: Leafcutter Ant

Leafcutter ant tending fungus

- Each member is a different size with a different look that is associated with the job they have. Most colonies have four castes.
- Mostly found in South America, Central America, and Mexico. Some are found in the southern U.S.
- After a few years, a colony can include 8 million individuals.
- They tend fungus for food in a relationship that has been going on for 50 million years.
- They can secrete an antibody that is the same one used to create the majority of antibiotics.

Part V The Real Bugs: True Bugs (Order Hemiptera)

While not the oldest evolutionarily, the order Hemiptera includes what entomologists call the "true bugs," the real ones, the OG. The word *bug* comes from the Middle English word *bugge* for "spirit" or "ghost."[65] And it makes sense: If you were to wake one morning with mysterious red dots with no warning, you might start to think they were put there by a visitation from a ghost (in this case, most likely a bed bug). And while we lay people use the word *bug* to mean lots of different animals, entomologists only attach it to this order.

True bugs' scientific name, *hemiptera*, means "half-winged". This is in reference to their wings that are hardened closer to their body but membranous on the second half. Not quite hard like the elytra of beetles but not completely membranous like bees. These wings lay flat across each of these bugs' backs or are kind of tented, such as a cicada's wings.

Another distinctive feature of this order? They all have stabbing mouthparts, which some use for plants and plant fluids and others use for sucking fluids from living creatures. For many of them, their mouthparts lie flat against the underside of their body until they are ready to use it. They rarely use this characteristic as a defense mechanism.

Some hemipterans, like cicadas and leafhoppers/treehoppers, appear in fossil records from 280 million years ago and are relatively similar in appearance to their ancestors. Many true bugs have the ability to produce loud noises (like cicadas), but a number of species are inaudible to humans without special equipment.

In terms of diet, some can be highly specialized and only feed on a single species of plant (which may lead to a higher chance of co-extinction), while others are more generalist feeders (or don't eat at all in their adult forms). They are also an important food source for many vertebrates like birds and lizards, as well as other invertebrates,

65 Schaefer, C. W. Prosorrhyncha. In Resh, V. H., & Cardé, R. T. (eds.), *Encyclopedia of Insects.* (2nd edition, pp. 839-855). Elsevier.

like spiders, assassin bugs (which are themselves true bugs), wasps (Eastern cicada killer), and robber flies.

Similar to beetles, some hemipterans are aquatic or semi-aquatic. The majority of species are found on flowers or plants, either for food or in order to catch prey. There are a few species who are known to be significant agricultural pests, not including aphids. A few planthoppers are known to cause serious destruction to crops like corn. They can be pesty in their adult and larval forms. Many of the larval forms develop underground so detection of them may be difficult.

However, scientists discourage the use of pesticides in this case because it not only leads to resistance in the bugs, but also harms everything that comes into contact with it. As an alternative to pesticides, many agricultural scientists are determining the effectiveness of parasitic wasps in reducing true bug populations.

True bugs have four suborders total, but one of them (suborder Coleorrhyncha) will not be discussed in this section . There are only about 30 known species in that suborder and all feed on moss. They are found in New Zealand, Australia, New Caledonia, and South America. The other three suborders are significantly larger and are found all around the globe.

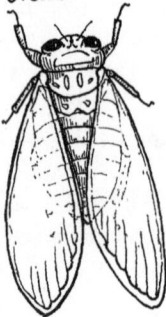
cicada

stink

assassin

leafhopper

Chapter 12: Aphids, Scale Insects (Suborder Sternorrhyncha)

Starbucks got in trouble in March 2012 for a surprising reason. People figured out that the company was using an insect as food dye in a number of their products, including anything with a red color. The cochineal bug has been used as a red dye for centuries (since 2000 BCE) and is approved by the FDA as a natural food coloring. People were upset that it wasn't vegan (even though I am unsure if that was claimed by Starbucks or not) or that it could be an allergen. The coffee shop defended itself, saying that the dye was legal, but by April of that year, they had decided to stop using it. Now they use an artificial red color. Tough question: natural but a squashed bug? Or artificial but not a bug?

Aphids and scale insects like the cochineal bug, along with a few others, are part of the suborder Sternorrhyncha. Humans have found many practical uses for true bugs, especially the ones in this suborder. We like bugs when they can be profitable for us.

All of the bugs in this suborder can be found on plants. They are often found in large groups or clusters. Some have wings, some don't, but all of them can travel best when carried by wind. Compared to some of the other members of their order, they are fairly small. Their size is definitely part of why wind travel is so successful for them. Of the thousands of species, only a few are considered pests and that is only when they are found in extremely high numbers.

How Aphids and Scale Insects Shape Our World

Scale insects produce a blood-like, red substance, called carmine, as a defense mechanism to ward off predators. They then suck the "blood" back in so as not to lose too many liquids. And that substance is a popular dye used by humans in various applications.

Adult female and male cochineal bugs look markedly different. Males have wings and slender bodies, while the females look more like an engorged tick. It is difficult to see the head or the body segments.

Females also don't have wings and for mating purposes, the males come to them. When the males and females of a species look drastically different, it's called sexual dimorphism. Like the difference between male and female peacocks. They are found on cacti and may actually just look like white dust. They don't move much so it's hard to tell that they are bugs. Rarely do they get to large enough numbers to cause significant damage to the plants they eat. Other animals, including other insects, eat them.

One true bug that grinds gardeners' gears is the aphid, what some may call the pestiest pest of the home garden. However, of the 4,300 species that exist, only about 100 of them are viewed as pests. They have been around for a long time too. They first showed up in the fossil record about 230 million years ago, about 50 million years after cicadas and leafhoppers first appeared. To other animals, they are not a pest. Some species of ant even keep them as cattle because they like the sugary liquid the aphids secrete. Lady bugs, particularly the nymphs, love aphids; not for their sugary goodness but as food. The nymphs are capable of eating up to 75 aphids a day.

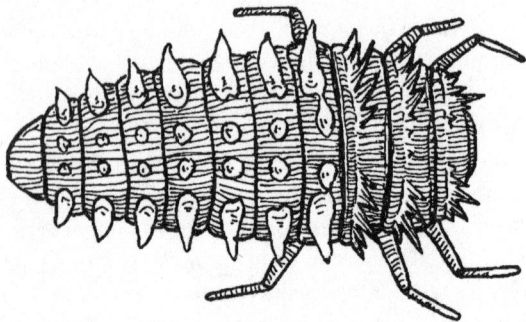

Young ladybugs look very different from adult forms and are often mistaken for something sinister. They are fairly small and can be found almost anywhere aphids are.

There are some species of wasp that parasitize aphids as well. All of the hemipterans in this group have parasitoid wasps that target them, turning them into living, walking buffets.

Aphid colonies are mostly female and they give birth to live young. They spread from plant to plant through wind. Even those species with wings travel better via air currents. They let go and get swept away to another plant. They love the color yellow and most often are attracted to plants with yellow flowers or leaves. Remember that even with a decent number of aphids on your plants, it may not cause serious issues. They need to eat too. Even though they were all over my sunflowers last year, the plants did okay and the longhorn bees loved them. I got flowers, and the bees, aphids, beetles, and ants got food. Everyone won.

Many aphid species give birth to live young. One species is able to carry their daughters in their reproductive tract, while their daughters are also carrying their daughters in their reproductive tract. Three generations, all in the mom/grandmother's body like a turducken. Depending on the species, they can have upwards of forty generations a year (which is how they can cause such havoc). They live in clusters, which I'm sure some of you have seen when turning over a leaf on your sunflowers. In personal gardens, some people use neem oil to discourage aphids from feeding on their plants. They like a lot of the decorative/pollinator plants that people put in their gardens.

Agriculturally, they can cause large issues for farmers. They were the cause of the Great French Wine Blight, where they decimated vineyards across Europe. The only way to fight them was to graft naturally resistant American vines (where the aphids are native) to more susceptible European species. The aphid is still being found in new parts of the world every year and vineyards keep trying new ways to eradicate them. At this point, there is no way to fully be rid of them and everyone has to rely on grafting with American vines. Eradicating insects does not often go well for humans, simply because, in the long run, we are unable to do so. Another story of non-native species introduction fucking stuff up, at no fault of their own.

The other bugs in this suborder have fewer species: whiteflies and jumping plant lice. They have similar lives to the larger families in this order. Whiteflies look like a mixture between moths and flies, but have a stabbing mouthpart and are hemipteran. Their wings have a white dusting, similar to the coloring of the cochineal bugs. You can often find them chilling, eating, and laying eggs on the underside of leaves. Like aphids, they also produce honeydew but don't seem to be attended by ants. Adults are able to fly short distances. They are known to be pests on certain plants, particularly in citrus orchards and some ornamental plants like orchids. Integrated Pest Management (IPM) uses parasitic wasps, lady bug larvae, and spreadable fungi to keep these populations low (it doesn't always work). They are still a great source of food for many small animals.

Jumping plant lice and whiteflies have been observed performing mating rituals by fluttering wings and creating sounds with their abdomen. Plant lice are not a type of lice, so if you see them on your plants, there is no need to worry about them spreading to humans. Most species are only found in the tropics. About 100 of the 2,500 can be found in North America. Like their family members, they use their stabbing mouthpart to suck nutrients out of plants. Plant lice are also citrus pests who can be parasitized by wasps or flies. As the plant lice grows, it creates a little structure for itself on leaves to keep itself safe. These structures are called "lerps" and they make the covering themselves. They look like an opaque waxy substance. Their pestiness is balanced by predators. All in balance.

How Humans Can Interact with Aphids and Scale Insects

In regards to aphids in your garden, remember Xerces' recommendation when you find them on your plants: how much damage is actually okay for my garden? Do the plants look okay and if so, can I just let everyone eat what they need too? Obviously, you do not have to follow this advice, so try natural remedies like neem oil, non-neonicotinoid sprays, and using your hand to pick off as many as you can. Also, keep your fingers crossed that lady bug larvae and parasitic wasps find their way to your garden.

Humans love the vibrant red color that comes from smashing cochineal bugs. Many places around the world (including the U.S.) still use it as a natural dye and food coloring. The FDA lists it as a usable natural dye. As mentioned earlier, Starbucks got in trouble for using it, but other businesses still use these bugs as a natural food coloring, without nearly the same scrutiny. The dye is still actively used (and described as Natural Red 4 or E-12) in makeup, meats, juices, and many other products in the U.S. alone.

In other parts of the world, carmine from cochineal bugs is used as a clothing dye and is particularly important to traditional Mexican attire. Scientists are trying to replicate the color in labs, but the cost to do it in large quantities has been unsustainable. In order to get a pound of cochineal bugs, you need about 70,000 of them. Those

70,000 only yield about a fifth of a pound of the color. A few studies have been done to try and replicate it. The bugs still just do it better and scientists haven't quite figured out why. For now, cochineal bugs are here to stay as an important red dye.

Lac bugs, another true bug important to humans and the economy, produce dye and lac resin because they secrete a substance called shellac. It is commonly used in India and Vietnam (where it has helped build back up the economy in impoverished areas). In India, 20,000 metric tons are made every year. The bugs are also used to create a red and/or yellowish dye. It is approved for use by the FDA on furniture, fruit, and pills. It's also available for use on floors, leather, paint, and quite a few other surfaces. The color does not yellow in sunlight. A recent study has shown that a certain component of lac dye possesses antineoplastic or anticancer effects, so we may be seeing a different use for lac bugs in the future.[66]

Lac bugs are also used in food. Confectioners glaze is a type of shellac.[67] It also takes huge numbers of lac bugs to create a small amount of the substance, estimated at hundreds of thousands of the bug. It can be found in candy corn as well as almost all candies with confectioner's glaze. You eat bugs all the time and don't even know it. The actual pieces of the bug are no longer in the glaze, but it comes from farmed bugs around the world.

How Human Culture Is Shaped by Aphids and Scale Insects

As I mentioned above, ants are known to "farm" aphids. Aphids (and some other scale insects) secrete a honeydew-like liquid from their abdomen that is a sweet treat for ants. Because of that, ants can be protectors for large numbers of aphids. In the movie *A Bug's Life*, the queen ant has a cute little aphid pet that she carries around with her like a miniature dog the entire film. Its name is Alphie. Normally, ants don't just have one aphid since they still need to be in groups to continue mating and growing the colony, which means more honeydew. At different life stages, some aphids are "soldiers" and are in charge of protecting the colony. Some aphids live in galls and when

66 Shamim, G., Ranjan, S. K., Pandey, D. M., Sharma, K. K., & Ramani, R. (2016). Lac dye as a potential anti-neoplastic agent. *Journal of Cancer Research and Therapeutics, 12*(2), 1033-1035.
67 MacNaugthen, W. (2019, Apr 18). *Forget the Easter bunny. Let's celebrate the Easter bug.* New York Times. nytimes.com/2019/04/18/business/lac-bug-candy.html.

a predator tries to get in, like a wasp, these soldier aphids explode to cover the hole made by the predator. As they die, they use their little legs to smooth out the liquid.[68] They sacrifice themselves for the group. Since they can have upwards of forty generations a year, they aren't sweating the loss of a few soldiers.

Quick Facts: Aphids and Scale Insects

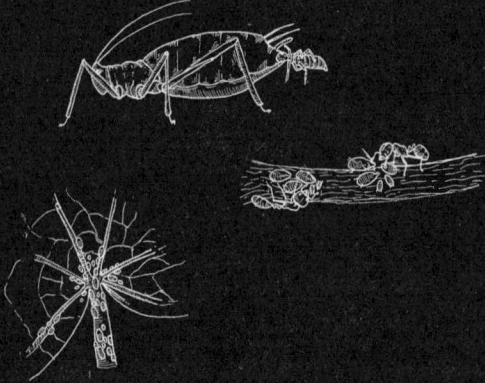

Top: Giving birth to live young. Middle: Ant farming aphids. Bottom: What one might see on the back of their plants should aphids be present.

- Have been used by humans for color dye and shellac for centuries.
- Aphids are often considered the ultimate pests of gardens.
- Many species give birth to live young.
- Often important food for many of the things that eat them, like ladybugs, lacewings, flies, and other insects. Farmed by ants for the nectar they produce, meaning they don't eat the bug but just the waste.
- Can travel via the wind to get to new places.

68 Kutsukake, M., Moriyama, M., Shigenobu, S., Meng, X. Y., Nikoh, N., Noda, C., ... & Fukatsu, T. (2019). Exaggeration and cooption of innate immunity for social defense. Proceedings of the National Academy of Sciences, 116(18), 8950-8959.

Bug Spotlight: Cochineal Bugs (*Dactylophius coccus*)

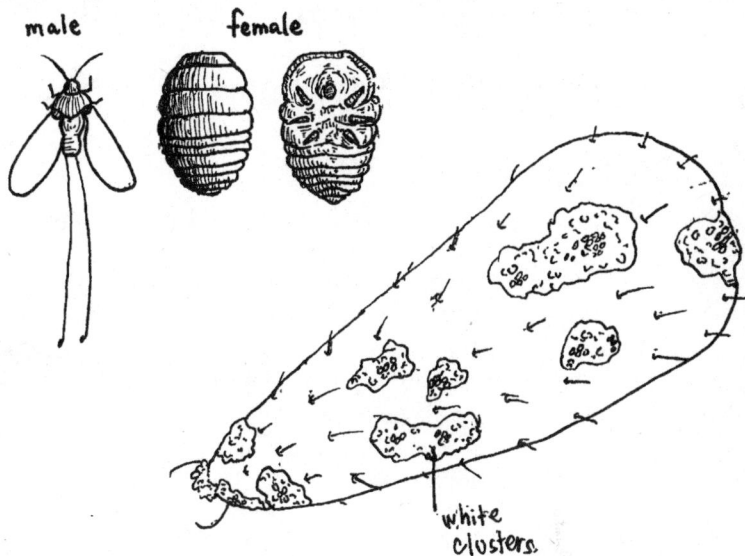

male   female

white clusters

Females, whose underside you rarely see, are about .2 inches in length, and males, who you rarely see anywhere, are about 1/32 of an inch.

- Used to dye food and clothing red (carmine, to be more specific), in the past and currently. Most often used in food and lipstick.
- Females do not have wings and will hang out on the same plant for their lives. Males fly around to find females.
- Often don't move and have wispy white hairs that make them look very un-buglike.
- Found in clusters on prickly pear cactuses. Humans have successfully farmed them in some parts of the world, but failed in others.
- It takes 70,000 individuals to make 1 pound of dye.

CHAPTER 13: CICADAS, TREEHOPPERS, AND LEAFHOPPERS (SUBORDER AUCHENORRHYNCHA)

*T*his suborder of hemipterans has the largest number of species, including cicadas, treehoppers, and leafhoppers. Cicadas are big and clumsy alien looking true bugs, while treehoppers and leafhoppers are most often very tiny.

Most people who live in areas with cicadas associate them with summer and are able to identify the sounds in the trees. By contrast, people often experience treehoppers by feeling something on their arm, seeing something green and tiny, and then watching it hop away to who knows where (they can jump extremely high for their size). Cicadas make a ton of noise while treehoppers don't. They are still part of the same suborder. Taxonomy is weird.

Cicadas have inspired art and music, while treehoppers have more recently become loved for their fancy hats. Both of them provide much needed balance to the food chain, with one being more recognizable than the other.

How Cicadas, Treehoppers, and Leafhoppers Shape Our World

When cicadas emerge, they look different from their adult forms and it is something of an event. They don't have wings and are a mud covered, brownish, very slow moving, plump thing. They emerge slowly from the roots of trees and need to climb onto the tree to break out of their larval form. The grips on their legs are impressively strong. I don't know if you've ever tried to remove a shell from a tree after they've transformed, but they stick to those trees pretty well. Once they climb onto the tree and are secured in a stable spot, the top of their larval shell splits and they molt out of the hole. They are particularly vulnerable at that point and often molt at night when the largest number of their predators are sleeping. It is straight out of an alien movie. They are initially white before their blood pumps through their bodies. Like butterflies, their wings are wrinkled and need to expand with blood before they can properly take off. Even with wings that look fully formed, they may need to take more time (upwards of hours) to be fully prepared to leave the tree in search of shelter and mates.

An annual cicada, which are often significantly larger than the ones who pupate underground for long periods of time.

Most of the cicada's life is spent underground, sucking on roots. Some emerge every year, while some take upwards of seventeen years before they emerge. In 2024, a thirteen-year cicada brood (what a specific geographical group of cicadas is called) and a seventeen-year brood emerged together. Southern Illinois was covered as they had the two broods overlapping.[69] Around fifteen states total saw the emergence of one (or both) groups. It was quite the spectacle. It is something that won't happen again for another 225 years so it is a once in a lifetime sight. This is the first time it has happened since 1803. Experts expect there to be millions, if not trillions, that emerge at the end of the spring/beginning of summer.

Scientists have discovered that in years with large numbers of cicadas, caterpillar species are less likely to be kept in check. To a bird, a fat cicada is tastier looking than a caterpillar so birds go for the hemipterans. This allows more caterpillars than usual to survive the year and eat all the plants they want. Many species of cicada emerge annually (tending to be larger in size compared to the thirteen and seventeen year ones). In years with annual, thirteen year, and seventeen year cicadas emerging at once, caterpillars are likely to have

---

69 Resnik, B. (2024, Jan 23). *Where billions of cicadas will emerge this spring (and over the next decade), in one map*. Vox. vox.com/science/24047261/cicada-brood-xix-xiii-19-13-map

a field day. Things balance out and the next year, caterpillar numbers will be smaller.

Scientists are still not 100 percent sure why broods come out when they do, although there are theories. Entomologists believe that once the ground temperature is at 64 degrees for around 14 days, cicadas emerge.[70] They don't have any definitive answer as to why they emerge on odd years. It may be so that other animals that eat them can't sync up with their cycle. It is assumed that they emerge together to overwhelm predators. Animals can't eat trillions of cicadas so their numbers will continue to stay high enough for them to lay eggs and be at similar numbers by the time they emerge seventeen years later.

Relatives of cicadas include treehoppers and planthoppers. Leafhoppers and treehoppers are in a different family than planthoppers (which does not make using their common name easy), but I'm going to talk about them as one group since they have many similarities. You've almost definitely seen one of these, even if you don't know that you have (unless you grew up in Antarctica—all other continents have them). Similar to cicadas, they look alien. They have fairly large eyes on the sides of their head and can have some interesting designs. Unlike cicadas, they are very tiny so it may be difficult to see their features. Like, smaller-than-a-mosquito tiny.

Of leafhoppers and treehoppers, many of the common species in the U.S. are green or brownish in order to blend in with the local flora. Tropical species are much more likely to be brightly colored. Also, unlike cicadas, they have phenomenal jumping capabilities. I have felt a small tickle on my arm to only look down and see a small treehopper. When I move to touch it, it jumps spectacularly high, especially for something so small. It's hard to track where they go once they jump since it is so far from where they started. There is a genus of treehopper (genus *Entylia*) that can jump 176 times its body length and higher than 47 times its body length. That would be like a 5 foot tall person jumping 880 feet away.[71] The longest recorded long jump in a human at this point is just shy of 30 feet. What a stark difference.

70 Fitzgerald, K. (2016, Mar 22). *How do cicadas know when to emerge from the ground?* Entomology Today. entomologytoday.org/2016/03/22/how-do-cicadas-know-when-to-emerge-from-the-ground/

71 Burrows, M. (2013). Jumping mechanisms of treehopper insects (Hemiptera, Auchenorrhyncha, Membracidae). *Journal of Experimental Biology*, 216(5), 788-799.

Since they are so much tinier than their cicada siblings, their defense strategies are a bit different. Clearly, jumping is an important part of their defenses. Their legs are very strong and they sometimes use those to stand their ground. Females guard their eggs and, in some species, remain with young throughout their development. They can buzz to scare away predators or they can use their strong legs to kick the would-be predator off of the leaf. They also defend themselves through mimicry, camouflage, and waxy substance creation. Some of them mimic ants, wasps, or other animals that their predators may not want to deal with. They also can have very bright warning colors (even though they would not be a poisonous meal). You may also see spikes or horns of some type that are supposed to discourage swallowing. Many species can also blend in very well with leaves, especially after they create small waxy houses on the underside.

To humans, the aspect of these creatures that often looks the most stylish is their little helmets or hats (definitely not the scientific term). I mentioned above that some of them use bright colors or spikes, but some of the adaptations are mind-boggling. Some have helmets that mimic bugs (like an ant or wasp) while others have ones that look like helicopter blades. Scientists have discovered that these head coverings are the remnants of a third pair of wings that they lost about 300 million years ago.[72] Now, they fashion them into these extraordinary head pieces.

Top left: *Bocydium globulare*, a Brazilian species. Top right: generic treehopper with no accouterments. Bottom left: *Cyphonia clavata*, an ant mimic. Bottom right: *Heteronotus vespiformis*, a wasp mimic

How Humans Can Interact with Cicadas, Treehoppers, and Leafhoppers

Lorde, a musician, used the sounds of cicadas in one of her albums for that exact reason. They can be painfully loud, especially in large

---

72 Pennisi, E. (2011, May 4). *Treehopper camouflage derives from ancestral wing*. Science. science.org/content/article/treehopper-camouflage-derives-ancestral-wing

numbers. Two species, the Greengrocer/Yellow Monday and the Double Drummer, produce a noise intensity in excess of 120 dB at close range. This is approaching the pain threshold of the human ear. In contrast, some small species have songs so high in pitch that the noise is beyond the range of our hearing. Even in species we can't hear, males crease their "ear" so that they won't be deafened by their own call. Only males can sing; females are incapable of doing so. They shelter in the trees and call to each other. Without shelter, mates, or proper places to molt, cicadas go extinct. This has already happened to two known broods: Brood XI (last seen in 1954) and Brood XXI (last seen in 1870).[73] Scientists believe this is because of a reduction in tree cover and forests with undisturbed ground.

While they only live two weeks as adults, they can live underground for up to seventeen years. Brood X, a seventeen year cicada brood, came out in Indiana in 2021 and basically covered everything. In Missouri during a large brood year, a frozen yogurt shop made shakes with them. Unsurprisingly, they got in trouble from the FDA. People really seemed to enjoy the shakes though and described the taste of the added cicadas as peanut-like.

Bug cicadas aren't the only impactful "invasion" of this order. Starting in 2015, the U.S. began to notice and became concerned about an "invasion" of the non-native treehopper, the spotted lanternfly. The insect is native to China and Vietnam where it has natural predators. In the U.S., however, it can proliferate quickly and is not as easily kept in check. The lanternflies are actually really cute and looking at them, you would never know they were destructive. They prefer a host tree that is non-native to the states but is often used ornamentally in yards, called the Tree of Heaven. Once they are established, they also like to eat soybeans, crab apple trees, and stone fruits, causing issues for farmers. So, between this predator-less invasive bug and rapid growing invasive tree, we are in for a bad time. They are regularly spotted on the east coast and are moving their way west. Entomologists tell people to kill them on sight, but there is so far no evidence that this practice reduces numbers. They said the same thing about Japanese beetles, and the beetles won that war.

73 University of Connecticut. (n.d.) *Broods*. Periodical Cicada Information Pages. cicadas. uconn.edu/broods/.

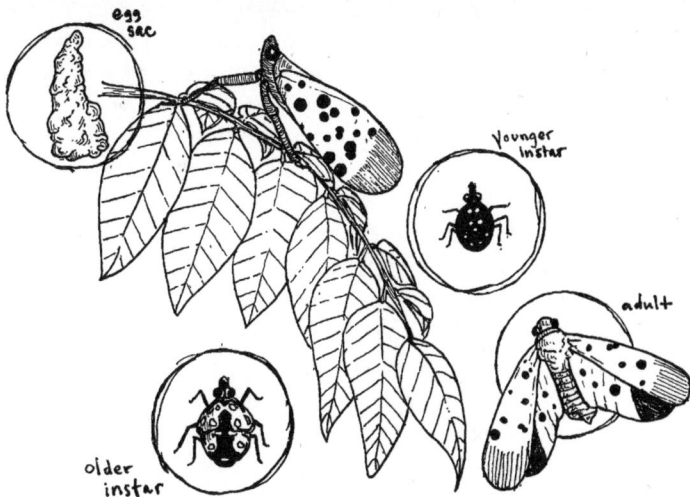

They are pretty though.

How Human Culture Is Shaped by Cicadas, Treehoppers, and Leafhoppers

Cicadas are frequently represented in human culture, particularly in Art Nouveau jewelry. The cicada, along with the dragonfly, is one of the most commonly found insects included in the style of that time period. The body of the actual cicada is not used, but their size, colors, and sounds are of interest to people. The annual cicadas can be startlingly large and can be brightly colored, so they can be inspiring for designs. In some cultures, cicadas represent rebirth and bring luck. Some of those same places eat them by frying them, sauteing them, or even eating them raw. During the 2021 Brood X emergence in the U.S., many news outlets posted stories about how to best cook them. According to *Popular Science*, the best time to eat them is right after they emerge from their exuvia (their shell that they shed).[74] If you are allergic to shellfish, DO NOT eat them, as you may have a similar allergic reaction when eating cicadas. The region of Provence in France has adopted the cicada as a mascot of sorts. They have a famous restaurant called La Cigale (the French word for cicada) but they do not have actual cicadas on the menu.

---

74 Gutierrez, S. (2022, Mar 9). The Brood X cicadas are coming, and you should eat them. Here's how. Popular Science. popsci.com/diy/cicada-cooking-guide/

Regardless of how long they spend underground, cicadas emerge with gorgeous wings that have antimicrobial properties and can also repel water. Scientists have been trying to harness the powers of cicada wings for various purposes. Since they naturally repel water, researchers have been trying to figure out how to make better fabric with similar qualities. They look at the cicada wings like a topographical map to determine its structures, to then determine how those structures create water and microbe resistance. They are thinking about how to use this research for everything from jackets to aircraft to medical equipment. Actual cicada wings will not be used, just manufactured pieces with the nanostructures of the wings.[75]

Surprisingly, cicadas are often mistaken for locusts, which they definitely are not. Bob Dylan's song "Day of the Locusts" is actually about cicadas. It may be because they can appear in very large numbers. Unlike locusts (grasshoppers really) and some relative true bugs, cicadas are unable to jump. They also don't eat as adults, instead relying exclusively on the food they consume underground as larva. Like locusts, they can fly but not nearly as gracefully. I've watched a cicada fly straight into a wall, fall down, right themselves, and then fly straight into the wall again. When you only live for two weeks and your sole purpose is to mate, you don't have a large brain.

Quick Facts: Cicadas, Treehoppers, and Planthoppers

- If you found a spotted lanternfly adult or larva, do not feel bad about smashing it. Also, let your local parks know so that they can keep track of numbers/sightings.
- Treehoppers once had a third set of wings that can now be seen on some species as an elaborate hat.
- All the insects in this order eat plant material by using their stabbing mouthpart.
- Planthoppers can jump 160 times their height; cicadas are not capable of jumping.
- Some cicadas are so loud, they can damage human hearing but most aren't. Planthoppers also sing but their song is too high pitched for humans to hear.

75 Román-Kustas, J., Hoffman, J. B., Reed, J. H., Gonsalves, A. E., Oh, J., Li, L., ... & Alleyne, M. (2020). Molecular and topographical organization: Influence on cicada wing wettability and bactericidal properties. *Advanced Materials Interfaces*, 7(10), 2000112.

Bug Spotlight: Seventeen Year Cicada (*Magicicada septendecim*)

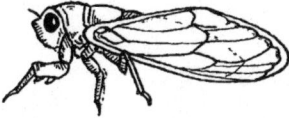

Left: Adult form of seventeen year cicada that are much smaller than annual species. Top right: The exuvia left behind after molting into an adult, often found on the trunks of trees. Bottom left: An example of the molting process which can look very alien. It's important not to touch them during this stage and let them do their thing.

- Called the Pharaoh cicada, only comes out once every seventeen years.
- Native to the U.S. and Canada.
- Their coloring looks different from annual cicadas: red eyes, bluish/blackish body, orange wings
- There are 12 broods in the U.S. that emerge different years. Brood XIV will emerge in 2025; Brood I will emerge in 2029.
- Two broods are extinct and others are at risk due to habitat development and lack of food.

CHAPTER 14: BEDBUGS, SHIELDBUGS, ASSASSIN BUGS (SUBORDER HETEROPTERA)

*T*his last suborder Heteroptera contains hemipterans with drastically different outward appearances and behaviors. Many of the species in this order do not eat plants but instead rely on animals as food. Most of them rely on other insects, while some feed on vertebrates. These bugs are seen frequently in urban spaces but are not known to the average person. I guarantee you've seen an assassin bug, even if you didn't realize what it was. They are a bit alien looking, with tiny heads and weirdly long legs on some. Many are aquatic or semi-aquatic. Some are huge and some are very tiny. Overall, they look quite different from the other suborders of hemipterans.

How Bedbugs, Shieldbugs, and Assassin Bugs Shape Our World

Since most are predaceous (feeding on live animal prey), they are less likely to be considered an agricultural pest and can be seen in a positive light. Assassin bugs are generalist feeders and will most likely attack anything that visits the plant they are on. They have been known to help with pest control. Many studies were done to determine if they could effectively be used in Integrated Pest Management, and it was found that the assassin bugs are full after only a few bugs. This means that they would need large numbers of assassins to make them effective. It isn't really feasible from a monetary or logistical standpoint. Some plant eaters have also been studied to determine their effectiveness on killing noxious plants. In Australia, they are still trying to control invasive prickly pear cactus using the cactus bug (which is native to the U.S.) and so far, the results aren't promising. They have found more luck using cochineal bugs.

Shield bugs, which are also sometimes called stink bugs, are a common sight across the globe: urban, rural, anywhere. They are called shield bugs (more common in England) because of the shape of their body. In the U.S., it is more common to call them stink bugs because of how they smell when smashed. The release of the smell is a defense mechanism to ward off predators. Some stink bugs also have bright colors to indicate how gross they would taste if eaten. Boxelder bugs, which are closely related to stink bugs, have a longer set of wings with bright red and black coloring (similar to a ladybug) to say "Don't do it. You'll regret it".

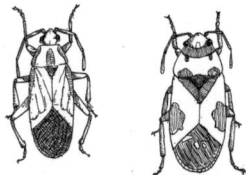

Left: Boxelder bug, *Boisea trivittata*. Right: Milkweed bug, *Lygaeus kalmii*. They look very similar but Boxelder bugs have orange outlines while the main color on a milkweed bug is orange with smaller black sections.

They carry similar markings to milkweed bugs (also hemipterans) who eat milkweed and are poisonous to predators. Boxelder bugs appear unappetizing to predators, but don't eat poisonous foods. If the predator wants to give it a try anyway, they release a stinky chemical to ward them off. So, milkweed bugs are poisonous, whereas boxelder bugs and shield bugs are just stinky.

Stink bugs have five segments on their antennae. Many insects are grouped based on the number of segments or type of formation they have on their antennae. They lay eggs on leaves of the plants that they eat, with the eggs in large clumps. When you see them hatch, they stay close together, both the newly hatched and parents. There are dozens of them, so people can be startled when seeing them. The brown marmorated stink bug is native to Asia but has found itself in the U.S. Since then, it has spread rapidly and since it has no natural predators, is doing pretty well to the chagrin of humans. They aren't particularly picky about what food they eat so they are known to feast on lots of different crops. Apparently, a newly introduced spider eats the shield bugs while other species seem to avoid it. Perhaps the spider can help wrangle the brown marmorated's numbers.

Assassin bugs are known to be great hunters and important for eating pests. They use their stabbing mouthpart to stab into their live prey, introduce a saliva that helps to break down the innards of that prey, and then suck it back into their bodies for nutrition. It is not advisable to pick one up since they have been known to defend themselves with their needle-sharp mouth. People have described the experience as unpleasant, which is unsurprising since you are getting stabbed. There are few types of assassin bugs: ambush bugs, kissing bugs, and wheel bugs. All of them, at least some of each genus, live in the U.S. Ambush bugs are so cool-looking. They have phenomenal camouflage and look very much like the plants they hang out on.

Ambush bugs will sit extremely still on a plant that they blend in with, waiting to catch something much larger than themselves. They have front legs that look more like a cicada nymph's legs than other assassin bugs. They use those legs to hold their prey in place. They are extremely patient. Sometimes their prey even walks right over them until they can get into a position that will more likely ensure their success in killing it.

Another type of assassin bug are the wheel bugs.

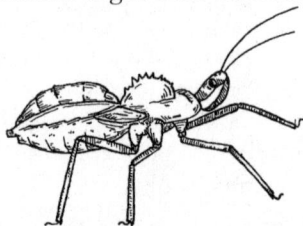

The North American wheel bug, *Arilus cristatus*, can grow up to 1.5 inches. Most similar species live only the Americas. They would prefer to be left alone, but if they feel threatened, their stabbing mouthparts can deliver a painful stab.

The North American wheel bug is a common species that can grow pretty big: upwards of 2 inches. I think they are super cool-looking, with a little half-wheel on its back. It loves eating all of the bugs that are considered pests: Japanese beetles, tent caterpillars, and dozens of other species that are considered detrimental. Since they eat all of these unliked bugs, they are considered beneficial insects. They do also eat good bugs, like lady beetles and honey/wild bees. Presence of wheel bugs in your garden is good. Dr. Raupp, an entomologist from University of Maryland, states that having them in your garden means that you have a healthy ecosystem in your yard.

How Human Can Interact with Bedbugs, Shieldbugs, and Assassin Bugs

The brown marmorated stink bug is an introduced species that is causing issues for farmers in the U.S. It is bothering small home gardens, as well as large agricultural farms. In 2018, it was estimated to have lost fruit farms, particularly apple and peach orchards, over $35 million. This little bug is no joke; however, another new accidentally introduced species to the U.S. is the Joro spider who is a natural predator of the stink bug. Scientists are hoping that they may help control numbers. Another strategy involves wasps. Can you guess what comes next? There is a species of wasp that parasitizes the brown marmorated stink bug. It is also from Asia and has been found in the U.S. Entomologists are trying to determine what the population size currently is and if they can be used to reduce numbers in the U.S. There is also a subfamily of stink bugs that are predaceous and have been known to feed on other stink bugs, just not the introduced brown marmorated bug. For other species, they are important for keeping those agricultural pests in check.

Kissing bugs are related to ambush and assassin bugs. Unlike the other members of the family, they are known to feed on vertebrate blood. Very few live in human dwellings, but there are a few species that are known to hang out in houses. They feed at night most often and are attracted to carbon dioxide, which we breathe out. When it comes to biting humans, they like to bite around the soft parts of the mouth and eyes (hence the kissing). In South, Central, and southern North America, kissing bugs are known to transmit the deadly Chagas disease. The bug bites you and then defecates on your face. When you scratch the bite, the poop can get into the bite, causing the transmission of the disease. Yes, I know that is all very unpleasant. It is not only found in humans but over a hundred other animals. If left untreated, it can be deadly but if caught early, it can be maintained. In the long term, even if people recover, they have a much higher chance of developing heart disease or heart-related issues (30 percent of people infected). Cases have been recorded in numerous U.S. states and, since the disease isn't common or easy to detect, scientists believe that only about 1 percent of the U.S. cases have been identified. New antibiotics are in testing phases and scientists will continue to research cures.[76]

Another set of vertebrate blood sucking bugs are bat bugs and bed bugs. Bat bugs feed exclusively on bats and look very similar to bed bugs. When everyone sees a bed bug in their home, they pray it's a bat

76 Andalo, P. (2023, Aug 24). Doctors advocate fresh efforts to combat Chagas disease, a silent killer. *CNN*. cnn.com/2023/08/24/health/chagas-disease-kff-health-news-partner/index.html.

bug (it usually isn't unfortunately). Bat bugs can feed on other animals if they are desperate but it doesn't sustain them for long. They truly depend on bats. Without them, bat bugs would cease to exist.

Bed bugs are a nuisance that have been plaguing humans for a long time. They are recorded in ancient Greek texts and the New Testament. We have shared space with them for centuries. It wasn't really until the 1980s that their spread became an acute global issue. Researchers believe that a lack of understanding of bed bug infestations, high mobility of humans across states and countries, and an overuse of underperforming chemicals increased the cases we have been seeing. In 2024, Paris experienced a bed bug "epidemic." What on earth is so scary about these little dudes?

The short answer is that they are difficult to completely eradicate. The long answer is longer. Hate them as much as you want, but they are evolutionary marvels. They have become so successful in the niche in which they reside that they are great at staying alive even when they are being systematically attacked. They can live in the tiniest crack in the wall, go almost a full year without eating, only feed once or twice a week, move only at night, and are almost impossible to smash. Their name comes from the fact that they are often found in the seams of mattresses or box springs. The saying "sleep tight, don't let the bed bugs bite" refers to them only coming out at night to feed. Unlike other relatives, they do not have wings, because they don't need them to be successful. They just run around instead of flying. Not having wings allows them to slip into very tiny spaces. I tried to smash one with my finger once and it did nothing. Little shits. They have become perfect at what they do.

As I stated in the introduction, having bed bugs can be traumatic. People experience serious psychological effects. I still check for them in the shower almost every day. You do not need to love them, but it is important to address the severe amount of misinformation surrounding them. Never do I expect you to be happy about them, but knowing a little more about them may assuage some fears. I love bugs but my heart still drops when I see a little fleck of something that is vaguely bedbug shaped. Learning about them helped me to understand them in a backwards sort of way.

How Human Culture Is Shaped by Bedbugs, Shieldbugs, and Assassin Bugs

In DC comics, there is a character named Ambush Bug who starts off as a villain, then becomes a hero. He has some Deadpool-like qualities in that he knows he's in a comic book. He was a fairly successful character and even had a few solo comic book arcs. He has appeared in DC comics as recently as 2011. In the popular Flashpoint series, Ambush Bug teams up with other DC insect-named super heroes, including Queen Bee, Firefly, and Blue Beetle, to take on the Amazons (Wonder Woman's people). Spoiler alert: They all die.

In earlier texts, some people believed bed bugs had medicinal value (that idea has been long abandoned). In the bible, scholars debate whether or not they are mentioned. At one point in the New Testament, one passage talks about bugs that bother someone in their bed in the middle of the night, that scurries back into the small joints of the bed. Probably bed bugs, but could be another night time menace. Regardless, they have been talked about and will continue to find themselves in our literature and stories.

Quick Facts: Bedbugs, Shieldbugs, and Assassin Bugs

- Assassin bugs, while perhaps a bit intimidating looking, are great friends to have in your garden as they are predators of "pests."
- Stink bugs (also known as shield bugs) are rarely a pest to people. but they do smell bad if smashed (hence the name).
- Their smell helps protect them from being eaten.
- The one caveat to the above statement is the brown marmorated stink bug. They are an introduced species that has no native predators. This is causing them to flourish with no checks and they are causing significant damage for farmers.
- Kissing bugs are not common in the U.S. but are starting to spread northward. The disease they spread, Chagas, is no joke. There are eleven species of kissing bug present in the U.S. and about 50 percent carry the disease. Texas, New Mexico, and Arizona have the largest number of sightings.[77]

77 Texas Ecological Laboratory (2024). Kissing Bugs & Chagas Disease in the United States, A Community Science Program. kissingbug.tamu.edu/

Bug Spotlight: Bedbugs (*Cimex lectularius*)

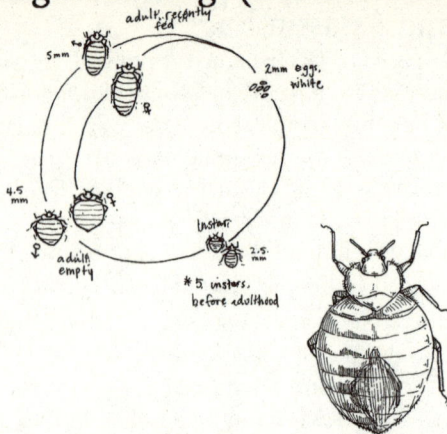

Most people who find them hope they are the closely-related bat bugs, which only feed on bats. You need a microscope to tell the difference. Unfortunately, they are often not bat bugs. Bed bugs go through incomplete metamorphosis, meaning the younger versions look like smaller versions of the adults.

- Can't fly but can run pretty quickly.
- You don't need to throw away all of your things if you get them.
- You need professional help to make sure that are completely gone. If you live in an apartment, treatment should be covered by the landlord.
- Anywhere can get bed bugs. Cleanliness has nothing to do with susceptibility of infestation.
- If you only find one, there are most likely many more. They are often found in beds or couches, but they can hide in a lot of different places, including the space between wooden floorboards or in photo frames if the infestation has gotten large.

Part VI: The Ones that Really Know How to Jump: Grasshoppers, Crickets, and Katydids (Order Orthopterans)

When I was a kid, one of my favorite books was *The Very Quiet Cricket*. Eric Carle does a great job in his books of making insects interesting to children. The little cricket held my heart more than some of Carle's other characters, like the hungry caterpillar or grouchy ladybug. Throughout the book, the cricket is trying to sing. He tries by rubbing his wings together, but he can't seem to make it work. All orthopterans, which includes grasshoppers, crickets, and katydids, are capable of making noise as a mating signal. Grasshoppers normally rub their legs together or along their abdomen. Crickets and katydids rub their wings together, which means that Eric Carle did his research (less so for *The Very Hungry Caterpillar*). The lovely, nighttime noises you hear outside your window in the summer are the mating calls of different orthopterans. They are also associated with quiet, like the term "quiet as a cricket," because of their abundant presence in the countryside. It is often used when someone does not respond to a question and there's silence: "It was crickets."

The bugs in this order, orthopterans (which means "straight wing") are known for their singing and their jumping. Their long wings lay flat across their backs. Think about a grasshopper: You don't see their wings until they take off, and when they do, their wings seem huge compared to their body. They all have large, exaggerated legs that are responsible for their ability to jump with force. Not all of them use their legs for such purposes though. A common defense mechanism for all species in this order is to kick their large, sometimes spiked legs, at their predator. If that doesn't work, they also are comfortable with losing a limb. The bottom part of their back leg pops off so that they can make their getaway, similar to a lizard losing its tail. Unlike lizards' tails, they don't grow back.

Most orthopteran species are masters of camouflage. Have you ever tried to find that cricket that you hear chirping? It's basically impossible, especially since they go silent when you get close. It has been the plot line in many TV shows where a character goes crazy trying to look for a cricket. Katydids are also known for their calls, but since they prefer trees to the space under windows in human homes, they don't get the same attention. Orthopterans show up in cultures around the world in various ways, most of the references are positive. There are a few particularly negative ones that have a hold on people (thanks, the bible), but these creatures are loved as food, story characters, folklore myths, and signaling summer.

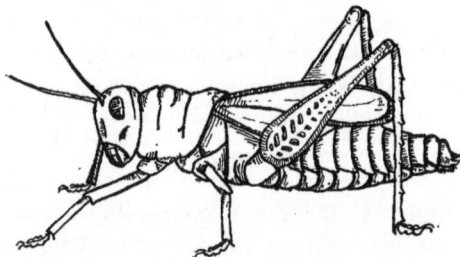

Katydid (top), cricket (middle), and grasshopper (bottom)

CHAPTER 15: CRICKETS (INFRAORDER GRYLLIDEA)

*C*rickets were my favorite bugs as a tween, and I used to catch them to keep as pets. They were all named Fred for some reason. Perhaps I wasn't very clever, but I wasn't alone in my love for them. Ralph Waldo Emerson once stated that if we could hear the moonlight, he believed it would sound like an intense, beautiful cricket song. Every cricket makes a species-specific song, which means there are over 900 distinct cricket sounds. They don't hear their songs in the same way we do; they don't have ears and instead hear with their legs. Some crickets (and katydids) chirp in relation to the temperature, speeding up or slowing down depending. The North American snowy tree cricket can be used to determine the temperature: number of chirps in 15 seconds + 40 degrees F should give you the approximate temperature. Sometimes it is difficult to single out one cricket since many of them sing simultaneously. Crickets are important to humans and may be the next big protein craze. They are already loved by many animals in the food web.

How Crickets Shape Our World

Many species of cricket are found in large numbers outside when the summer has been wet and warm. The rainfall helps with an increase in food available to them (plants) so they can produce in larger numbers. In drier seasons, you see fewer crickets simply because the food supply is not as ample (it may cause the opposite reaction in locusts). Large numbers are common in waves, and throughout history, crickets appear in large numbers in summer every once in a while. They don't damage crops or harm people, nor do they carry disease or bite. They are pretty chill and seeing these booms in their numbers is normal and a good sign. When we stop seeing large numbers every once in a while, that is when we need to worry.

With scientists discussing the impact of global warming and the effects the meat industry plays in the overall issue, people have been trying to find a protein source with a smaller carbon footprint. Enter crickets. Flour made from crickets is already being used as an additive in foods around the world to battle food and nutrition security. Their protein levels are similar to larger animals and they are full of the necessary vitamins. They also take up significantly less space and require less food. Their carbon footprint is significantly smaller than the current meat industry.

Currently, the cost of cricket flour is fairly high since it is still a pretty niche market. It also takes between 4,000 and 5,000 crickets to make a pound of flour. The insects are killed by going into cold temperatures which puts them in stasis. While in stasis (a normal response for them in cold temps), they are flash frozen before being ground into a fine powder.[78] The powder can be used to make bread, pasta, and any other food that is made with flour. The flavor is earthier than wheat flour.

How Humans Can Interact with Crickets

People all over the world have eaten crickets for centuries. Entomophagy (the act of eating insects as food) is common in many places but less so in the U.S. I went to the Insect Festival in Tucson many times. Each year, they had snacks to try that were made with insects. Over time, the snacks tasted better and better. Strides are being made in the cricket flour world, and it may reach a point where you cannot taste the difference between regular flour and cricket flour. Even as a bug lover and lover of the idea of entomophagy, it was slightly difficult to eat it knowing that it was made from crickets. My friend gagged on the first treat and refused to try the rest. The taste wasn't bad, just the knowledge of it being a cricket is what did it. Some animal rights advocates do not agree that mass killing of crickets is any better than the current meat industry. There is a growing field of insect welfare.

Insects are often viewed differently from other animals. Research shows that they, especially in media, are not described as sentient animals, which is a category reserved for vertebrates.[79] Insects are also often described as "harvested" like a plant, rather than killed like an animal. In educational interpretation at zoos, they discourage educators from anthropomorphizing most animals, but encourage the opposite for insects. Giving crickets a more human-like story reduces the gap between us and them. Think about how bad you might feel if you were to eat Jiminy Cricket. It worked to create more interest and

78 Cricket Flours. (n.d.) FAQ. cricketflours.com/faq/

79 Delvendahl, N., Rumpold, B. A., & Langen, N. (2022). Edible insects as food—insect welfare and ethical aspects from a consumer perspective. Insects, 13(2), 121.

care in honeybees during colony collapse: they were described as hard workers who were being killed. Using those tactics in the media may shift people's opinions of eating crickets as food.

There will always be a contingent of people who will not agree with the farming of insects as food (like vegans or organizations like PETA) but that doesn't stop most people from viewing it as an alternative. On the other end of the spectrum are people who view cricket protein as addressing global food and nutrition insecurity, who may understand the difficult ethics but have made the decision to focus on human health more than the death of insects.[80] Even without the animal welfare protections, the facilities that farm crickets are held to the normal health standards for other farms. Commercial farms need to successfully pass inspection by the USDA's Federal Safety and Inspection Services in order to operate.

The physical appearance of some crickets can make them difficult to identify. A few crickets are particularly strange looking and are often mistaken for another insect or some alien creature. Looking at some of these crickets, it is hard to determine just why they all look so drastically different from each other, even if they are all within the same order. For example, the mole cricket developed front legs unlike any other insects (but surprisingly similar to vertebrate moles) in order to be better at digging. They have the ability to fly as adults, but they live most of their lives burrowed underground. Some species have been introduced to countries where they have no natural predators so their burrowing can be considered pest-ish.[81]

Jerusalem crickets aren't even technically crickets (as the name would suggest) but are in the order Orthoptera so closely related. They are more closely tied to wētās, a giant native New Zealand orthoptera (also not a cricket but has cricket features). The director Peter Jackson named his graphics company after the wētā. Orthopterans in this category are flightless and have large mandibles that can give

---

80 Magara, H. J. O., Niassy, S., Ayieko, M. A., Mukundamago, M., Egonyu, J. P., Tanga, C. M., Kimathi, E. K., Ongere, J. O., Fiaboe, K. K. M., Hugel, S., Orinda, M. A., Roos, N., & Ekesi, S. (2021). Edible Crickets (Orthoptera) Around the World: Distribution, Nutritional Value, and Other Benefits-A Review. Frontiers in nutrition, 7, 537915. doi.org/10.3389/fnut.2020.537915

81 NOTE: As you may have noticed, I have said this in every chapter and will continue to say this in every chapter. Quite pest species are introduced/non-native species that have no native predators.

a seriously unpleasant bite. They also can release a foul-smelling chemical.

Cave crickets (which are true crickets who are also sometimes called camel or spider crickets) have a strange humped appearance. They have such long antennae and legs that extend much higher than their body. They live most of their lives in caves so they use their antennae and legs to feel around in the dark. You may find some in your basement as they like damp, dark places. They are not dangerous, nor are they a pest so can easily be moved outside.

How Human Culture Is Shaped by Crickets

With the complete eclipse in April of 2024, NASA put out a general call to the public to record sounds of crickets and birds during the event. A group of citizen scientists (regular people participating in gathering scientific evidence) did a similar thing during the eclipse in 1923. Scientists have discovered that nocturnal animals during an eclipse can get confused and you may hear some nighttime noises that you would not normally hear at noon. NASA wants to capture the sounds of the eclipse to study how the creatures of earth respond to the event. If you want to listen to sounds, you can visit the official NASA citizen science page: eclipsesoundscapes.org/.[82]

In human history, crickets are involved in an important milestone: The first example of an insect drawn by humans was a cricket carved into a bison bone about 14,000 years ago. They are still being drawn by humans today but in very different mediums. Jiminy Cricket was an important Disney figure in the 1950s and 60s. He still shows up from time to time. He was an anthropomorphized version of a cricket, meaning that he was made to look more human so that audiences connected with him. In Mulan, her cricket, Cri-kee, is more physically accurate: six legs, doesn't walk on hind legs, is not in clothes, doesn't talk, etc., all the things Jiminy had. Both Jiminy and Cri-kee could possibly be described as "merry." The term "merry as a cricket" comes from their constant singing, once again an anthropomorphized interpretation of that behavior. Their songs are meant to attract a mate, not an expression of their feelings. Crickets, or at least their "merry" sounds, are easily recognizable.

82 Paul, A. (2024, Feb 23). NASA wants you to record crickets during April's solar eclipse. Popular Science. popsci.com/science/nasa-eclipse-study-soundscapes/

Interestingly, they are also used for fighting in Asia. Humans bet on cricket fights and the winner is the one who kills or pulls off the hind legs of their competitor. The crickets are worked into a fury using pheromones from a female who is ready to mate, which turns the males into crazed fighters. Even though they can be made to fight in the ring, they are not known to often fight that viciously in the wild. At least, there are not a lot of documented cases outside of the ring or the lab, apart from light charging or tussling. Humans have a history of keeping crickets that dates back to the eleventh century. In Japan and China, they were considered good luck and often kept for their nighttime singing. In the video game *Ghosts of Tsushima*, you can collect crickets as a side quest. It is a lovely way to gain experience and listen to cricket song. Win-win.

Quick Facts: Crickets

- A cricket was the first animal carved into bone by distant humans.
- Their chirps can help determine the temperature: however many times they chirp in 15 seconds + 40 should give you the degrees in Fahrenheit.
- Crickets may be the protein of the future. They are able to be kept in much smaller spaces with less food and water. They give off less CO_2 and have more protein per serving than meat.
- Crickets are already a very important food source in the food web
- Seeing one in your house is not an indication of an infestation. They don't really do that. Nor do they bite, so if you find one in your home, just put it back outside.

Bug Spotlight: Strange Looking Crickets

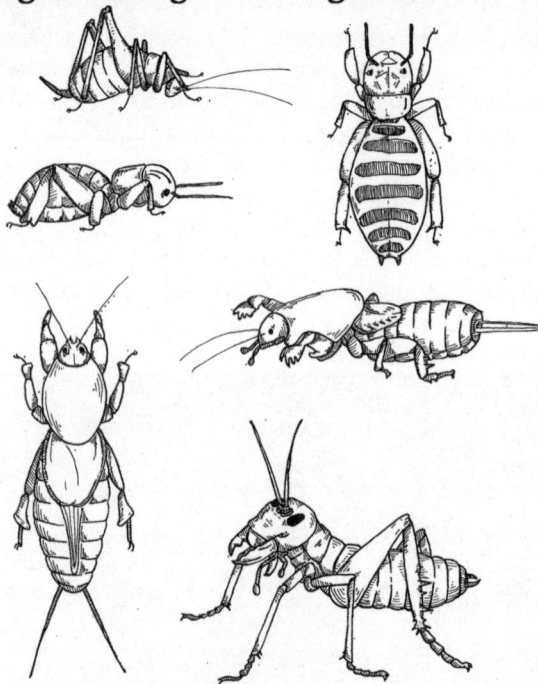

Top left: Giant cave cricket. Top right and beneath top left: Jerusalem cricket (an orthopteran but not a cricket, even though it resembles one). Bottom left and beneath top right: mole cricket. Bottom right: weta (an orthopteran but not a cricket, even though it resembles one.)

- Jerusalem Cricket: In Orthoptera, but not actually a cricket; has pretty strong jaw, so watch out for a bite. Doesn't have wings and can't jump—so easy to push into a cup—around 1 to 2 inches.

- Mole cricket: Similar front feet to a mole. They do burrow underground. Can't jump. Around 1 to 2 inches.

- Weta: Found only in New Zealand. Doesn't have wings and is closely related to Jerusalem cricket. Can be from 2 to 4 inches.

- Cave cricket: Like damp dark places best. May end in your basement for that reason, but they are harmless. Pretty good jumpers. Often have a large ovipositor that may look like stingers (100 percent not though.) Can be .5 to 2 inches.

CHAPTER 16: KATYDIDS (INFRAORDER TETTIGONIIDEA)

*T*f you've ever seen a walking leaf in the U.S.A., chances are it was a katydid. Some people call them "bush crickets," mostly outside the U.S. They are often found in bushes and trees, which differs from their relatives. While in the trees, it is almost impossible to spot them. They also will move slowly or not at all during the day to camouflage themselves (a form of crypsis, which means concealing itself). They look exactly like a leaf, including the veins. Similar to crickets, they can help you determine the temperature outside: number of chirps in 15 seconds + 37 degrees F. Most chirps are used as mating calls, but some are for more sinister reasons. A species in Australia, the spotted predatory katydid, mimics the sound of a native female cicada's wing movements. They do this to lure the male cicada to them for a meal.[83] Just when you think you have bugs figured out, they throw in a villainous curveball.

How Katydids Shape Our World

Katydids can be vegetarian or predatory, with the majority being omnivorous. Most often, katydids are greenish or brownish, sometimes yellowish; basically summer leaf colors; however, like other species, they can be born a different color based on a recessive gene trait, although this is rare. Some animals are sometimes born with albinism (no pigment). Katydids, however, can be born with erythrism. Erythrism is also found in other animals, and it causes the body of the animal to have a reddish pink hue. In katydids, they are bright pink. Many of these insects do not survive into adulthood since their ability to camouflage is impossible. People have been recording pictures of them a lot recently and I would recommend looking them up. They look adorable to us but like a tasty snack for other animals. Even if they are their normal camouflage color, moving around is dangerous for them. You can trick someone by looking like a leaf but when you start walking around, the jig is up. Females are known to move around more than males. They move to find a mate(s) or in order to lay their eggs in several places. Females are likely to be eaten in larger numbers

83 Marshall, D. C., & Hill, K. B. (2009). Versatile aggressive mimicry of cicadas by an Australian predatory katydid. PloS one, 4(1), e4185. doi.org/10.1371/journal.pone.0004185

than males because of that. Some scientists believe that larger numbers of females are born each year to account for the larger number of female deaths.[84]

From the perspective of them as prey, they have some pretty good defense mechanisms. I'm not sure if you've ever seen one take off, but they can be surprising and disorienting. They use their quick, large wings to befuddle the predator and give them time to escape. Some have pretty sharp spikes on their legs, similar to grasshoppers, that they can use to basically kick-stab whoever is bothering them. That doesn't work against humans unfortunately, so they find themselves on the menu. In Malaysia, they have been eating katydids and other orthopterans for centuries. The queen herself used to love them soaked in salt water for thirty minutes and then fried in fat. People still eat them regularly but it is made at home, not typically offered at restaurants. They can be found as street food and are an important source of income for some people. Katydids are not farmed but caught in the wild. Children, most often, will go collecting katydids at night (since most are nocturnal) to eat for snacks or the main course the next day. After soaking, they can be tossed in different spices and fried to improve the flavor. Similar to crickets, researchers are trying to determine if katydids can help with food insecure nations by using the katydids and other relatives almost as a second crop: first, eat the crops, then eat the things that eat the crops.

Katydids have been around for quite some time and scientists are still discovering fossilized specimens or specimens encased in amber. In 2023, two important papers were published about new information that 50-million-year-old katydids had revealed. The first one was an actual fossil, which showed some extremely detailed information about the insect. Insect fossils have been found before but are found much less often because the insect's soft body disappears before the fossilization process can happen. Fossilization is much more likely to happen if the animal has bones. This katydid fossil preserved some internal organs that surprised the researchers. The testes of the male were preserved as were some other testes-related structures. They compared the

---

84 Note: there is the cutest video of a pygmy marmoset encountering a katydid for the first time. The bug doesn't move very much, but every time it does, the monkey looks at it with confusion or disgust? I feel like I'm anthropomorphizing it. It reaches out to touch it but is so shocked when it moves. It's amazing.

example to an extant species in the same genus and found remarkable similarities.[85]

The second paper was based on information found from a katydid in amber. The specimen, which was about 44 million years old, captured an important time in the evolution of katydid as a family. Katydids (and moths) have been in an "evolutionary arms race" with bats throughout their time on earth. Katydids are able to hear the bat's use of echolocation and have been observed going quiet when bats are nearby. Bats can hear the mating calls of katydids. In this particular specimen, it showed a period of time when the katydid had evolved to have mating calls that were outside of the hearing levels of bats. They evolved this to avoid detection. Bats and katydids have continued to evolve different hearing and location abilities in their predator-prey relationship.

You may have also seen one or two with what looks like a giant hook coming out the back, maybe like a stinger. That apparatus is not a stinger but an ovipositor. First, that means you saw a female katydid. Second, that hook is used to lay eggs and is not used for anything sinister (similar to ichneumon wasps). The ovipositor can become bigger or smaller and does look strange when they are depositing eggs. You can see the body moving in a way that may look like it is going to sting, but it can't. It's just taking the time to lay some eggs in a tree. I've even seen them do it in the cracks of a sidewalk. Katydids are incapable of stinging, but they have been known to bite, especially if you have them enclosed in your hand. Their bite is more surprising than anything else and doesn't really hurt. They really just don't want to have to deal with you because you are way bigger than they are.

When they mate, the male brings a gift to the female to eat which provides necessary nutrients to her. She then also eats the gift instead of his sperm so the male is more likely to fertilize the eggs after giving a gift. They are polyamorous (which is very common in insects) in the sense that both males and females mate with numerous partners. The first male that the katydid mates with is most likely to be the father, if the female mates with a second partner soon afterwards. So one sperm packet pushes the other sperm packet further in, making

85 Heads, S. W., Thomas, M. J., Hedlund, T. J., & Wang, Y. (2023). A new fossil katydid of the genus Arethaea Stål (Orthoptera: Tettigoniidae) with exceptionally preserved internal organs from the Eocene Green River Formation of Colorado. Palaeoentomology, 6(3), 268-277.

the fertilization more likely. Some biologists believe that polyamory is more common in species that don't provide parental care. Katydids lay their fertilized eggs and do not stick around to help raise or feed them (extremely common in the insect world).

How Humans Can Interact with Katydids

Outside of their ecological importance, katydids are important to cultures in different ways. In sub-Saharan African, there is a species that is used to treat fungal infections on the feet (Athlete's foot). Some species are used to lull customers to booths or stalls in markets. People are attracted to the sound of the katydid and are more likely to look their way. They are used in nature-sounds sleep machines, along with the sound of crickets. You can tell the difference when they are outside because the sound comes from above, not below. With a sound machine (or when in the market), it may be difficult to tell the difference between a katydid song and cricket chirp.

How Human Culture Is Shaped by Katydids

In English folklore, katydids were believed to carry the soul of a young woman or girl. The girl's soul was trapped in the insect because they had committed a murder. While in this form, they debate over whether or not they committed the murder for the rest of their existence: Katy Did or Katy Didn't. What a horrifying thought. Their name comes from the sound that many European species make. Not all species make this sound, however, and can often be confused with the sounds of cicadas and crickets. Again, determining who is making the sound is difficult since you can never seem to find them in the trees, and all of those insects are singing around the same time of year. Similar to crickets, each species has their own call so as not to confuse mates (or purposefully confuse them like the spotted predatory katydid in Australia).

Did Katy do it or didn't she?

Katydids don't show up in literature and art as much as other insects, but they have some representation. There are a few gorgeous poems about them, like the poem by Oliver Wendell Holmes simply titled "Katydid." He gets some things about them wrong (i.e., songs are not sung by females frequently), but it is a lovely poem. He is referencing the folklore of the young woman being turned into a katydid. There are also numerous kid books with titular characters named Katy that tell stories of the insects. The artist Charles Burchfield's later work focused on visualizing insect sounds, including many orthopterans. Undoubtedly, humans will continue to be fascinated and inspired by these walking leaves.

Quick Facts: Katydids

- Can be used to determine temperature with the same equation as crickets: how many times they chirp in 15 seconds + 40 degrees.
- They can sense bat echolocation and stop making mating calls when bats are around.
- Folklore suggested that katydids were the transformed bodies of a murderous woman named Katy, who was cursed to ask "Katy Did It, Katy Didn't" for eternity.
- Their noises, along with crickets, are associated with summer time and can be calming or comforting sounds for many.
- If you find a pink one, it may be pretty, but will most likely get eaten before others. They rely on looking like a leaf for camouflage to keep predators away. Leaves aren't pink often, so they stand out.

Bug Spotlight: Ovipositors

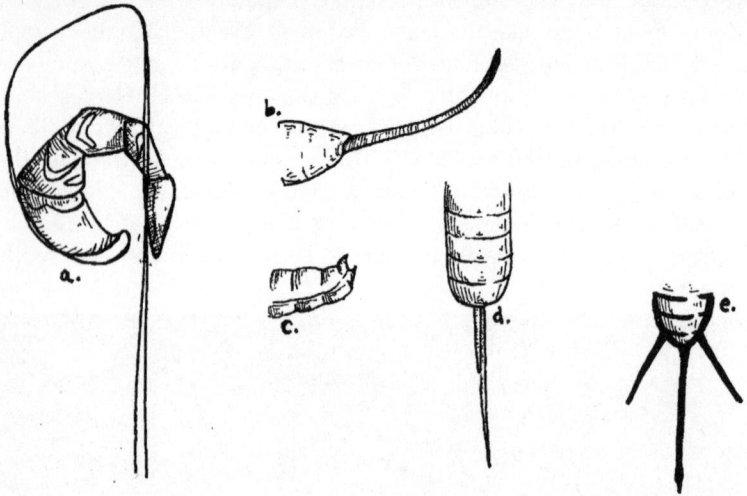

Various ovipositors found in insects. A. Giant ichneumon wasp. B. Katydid. C. Grasshopper. D. Horntail wasp. E. Cricket.

- Look like stingers, but they absolutely are not.
- Many ovipositors are flexible, like ichneumon ovipositors. This allows the female to accurately control where the eggs go.
- Some katydids have large ones that are less flexible but serve the same purpose.
- The end of the ovipositor is sharp enough or thin enough to get into, be it tree or sidewalk cracks, a space where the eggs will be safe.
- Almost every insect that has an ovipositor is incapable of stinging. A stinger is a modified ovipositor.

CHAPTER 17: GRASSHOPPERS (INFRAORDER ACRIDIDEA)

Grasshoppers are often easy to identify because of their large size, buff legs, and impressive jumping abilities. They can jump significantly higher than crickets and katydids. They are not poisonous or dangerous (they have been known to bite) but have been the antagonist in quite a few stories. Scientists fear a steep decline in their numbers. Grasshoppers have been around for a lot longer than humans so thinking of there being so few is troubling. The first fossilized relatives date back 300 to 250 million years. They've survived three of the big six extinctions, but we will need to see if they make it through the current (human-caused) event. While scientists have found multiple fossils of grown orthopterans, it was not until 2024 that the first grasshopper egg cluster was identified. Researchers in Oregon, who found it in 2012 originally, thought it was an ant egg cluster, but when they finally were able to get to the piece, they realized they were wrong. Being the first of its kind to be found, it will have significant importance in paleoentomology (which is the study of extinct bugs).[86]

How Grasshoppers Shape Our World

Grasshoppers are rarely in large enough numbers to be considered a pest. Anyone who has walked through a sunny field may be surprised to learn that, as they jump and glide from under your footsteps. As mentioned in the intro to orthopterans, their large straight wings can be surprising when extended since they blend in so well with the rest of their body. They are known to kick any potential predators as a deterrent but they also will often have bright flashes of color on the underside of their wings. When they take off, the bright color and sound of them flying away scares the would-be predator. They mostly just hang out, until they have to fly away, and are not a threat to you or your plants. That is until they become locusts. There is some general confusion about the difference between grasshoppers and locusts.

86 Sadiq, S. (2024, Jan 22). World's first known fossil of grasshopper eggs discovered in Eastern Oregon. *Oregon Public Broadcast*, oph org/article/2024/01/22/think-out-loud-first-known-fossil-grasshopper-eggs-discovered-eastern-oregon/.

To be fair, the information often presented to the public describes them as two distinct species. What is bonkers is that locusts are a type of grasshopper who goes through physiological changes as a reaction to their environment: They aren't born as locusts but become them. The grasshopper is a grasshopper, doing grasshopper stuff. Then, conditions that favor successful breeding exist, leading them to produce more in a concentrated area. This leads them to become locusts and their body goes through changes: adapted behavior and physiology including changing colors to brighter ones. In their new form, they swarm spontaneously. These behaviors are generally dependent on a drying out of the habitat following good breeding conditions.[87] They make their way to a food source that ultimately wipes out the crops but is enough to feed them. "A plague of locusts" as written about in the bible is a scary sight for those whose food source is being diminished. So, all locusts are grasshoppers but not all grasshoppers are locusts.

There are no species in the U.S. that can transform into locusts, at least not anymore. There is an extinct species called the rocky mountain locust. It lived across the U.S. and into southern Canada. It wasn't always in its locust form but would sometimes swarm and decimate farms during years of drought. During the early nineteenth century, they were common. Then suddenly, at the end of the nineteenth century they seemed to completely disappear. They had previously caused a serious famine in the 1870s which prompted the U.S. to create an entomology commission to determine how to best deal with the insects. The committee did research on the behavior and ecology of the locust, as well as created defensive strategies for farmers. When the next drought arrived however, the locusts were gone. As in, extinct.[88] Scientists were baffled. There are a few theories as to what happened. Jeff Lockwood, an entomologist who wrote a book on the rocky mountain locust, believes that their extinction was tied to habitat destruction. Much of the land where the locusts lived was being developed and disrupted an entire year's worth of eggs, essentially destroying what would be the next generation. Habitat destruction is one of the common underlying causes of many extinct, endangered, or vulnerable species currently.

87 Chapman, R. F. (2009). Locusts. In Resh, V. H., & Cardé, R. T. (eds.), *Encyclopedia of Insects.* (2nd edition, pp. 589-592). Elsevier.
88 Hopkins, T. L. (2005). Extinction of the rocky mountain locust. BioScience, 55(1), 80-82.

The difference between large numbers of grasshoppers swarming and a locust swarm is very different; however, we can still see large numbers of grasshoppers that can cause agricultural damage. These large numbers are not a swarm. Locust swarms can reach absurd numbers in the range of two to thirty billion species. Grasshopper swarms do not come anywhere close. Large amounts of precipitation can increase the number of eggs laid and food for them. This behavior is similar to other orthopterans. In other countries with species that can turn into locusts, they may flourish in the early season precipitation, but if arid conditions are created, they may change. Ecologically, grasshoppers are an important food source for animals, including humans.

How Humans Can Interact with Grasshoppers

Southern Mexico, and now other parts of Mexico, have used grasshoppers in their food for centuries. The Mexican word (and it is a word that is exclusive to Mexico and Central America) is *chapulines*, from the Nahuatl/Aztec word for food. Grasshoppers can be eaten as snacks or as additions to other dishes. I had a friend who went to Mexico City where they got guacamole. When it came to the table, it was covered in fried grasshoppers. They said it was delicious. In Seattle the baseball stadium started selling *chapulines* covered in a chili-lime seasoning at Mariner baseball games and people seem to enjoy them. This dish is actually one of the best-selling concessions.[89]

In Uganda, people love consuming grasshoppers like their Mexican mates but are lamenting the dwindling numbers. They commonly eat grasshoppers during the time of year they appear. People who rely on grasshoppers for extra money or full-time work are coming up short handed. Over the last thirty years, Uganda has lost a third of its forests. Workers who could easily fill eighty sacks in the past can only fill one now, which means a lot of lost money and work for people all the way down the line: those who catch, those who clean and prepare, and those who sell.[90] Not only does it mean less work and money, but also less food for those who rely on alternate (cheaper) forms of protein.

89 Kramer, D. (2021, Jan 13). Mariners' famed park food: Grasshoppers? Major League Baseball. mlb.com/news/mariners-fried-grasshoppers-backstory.
90 The Guardian. (n.d.). Where have all the grasshoppers gone? Uganda's insect traders struggle to find protein-rich bugs. theguardian.com/global-development/2023/dec/07/where-have-all-the-grasshoppers-gone-ugandas-insect-traders-struggle-to-find-protein-rich-bugs.

A group of scientists is studying the viability of using grasshoppers to detect explosives. At Washington University of Saint Louis, engineers attached electrodes to the brains of grasshoppers. Their sense of smell is complex and they are able to identify explosives by their smell with 60 percent accuracy. Sadly, it renders them paralyzed and they die of exhaustion after about seven hours.[91] Much less cool ending to the story. There is less concern about their livelihood since it's just an insect, right? Unsurprisingly, the project was funded by the Navy. The ethics around this are similar to the concern around farming crickets. Is this science okay because it is used on animals that aren't considered sentient?

A 2022 study found that there is either strong or substantial evidence that many orders of insects feel pain.[92] This hasn't been something often thought about in the past. I know that some people use orthopterans when going fishing. They just attach a live one to their hook by stabbing it right through the middle. Its movements are what attract the fish. This is a bigger question for research and farming since other animals have welfare guidelines. While the cleanliness of the farm is held to current cattle industry standards, those welfare guidelines do not currently apply to insects. Most animals under the Animal Welfare Act have guidelines for reducing pain/suffering in the killing of the animals. Insects, traditionally thought to not feel, are not included. The big question now is how to address the increasing number of farmed insects, insects used in scientific studies, and the evidence that supports the fact that they do feel pain.

On the other end of the spectrum, some people take such personalized care of their insects. At the Houston Zoo, keepers noticed that one of their grasshoppers after a molt seemed to be having trouble holding up her head. In order to help, they used a q-tip to create a neck brace for her. They left it on for a few days while her exoskeleton hardened. When they removed the neck brace, she no longer had trouble with the weight of her head. The species lives for a long time (in the context of insect longevity). In the wild, females

91 E+T Editorial Team. (2020, Feb 18). 'Cyborg' grasshopper engineered to sniff explosives. Engineering and Technology. eandt.theiet.org/2020/02/18/cyborg-grasshopper-engineered-sniff-explosives.

92 Gibbons, M., Crump, A., Barrett, M., Sarlak, S., Birch, J., & Chittka, L. (2022). Can insects feel pain? A review of the neural and behavioural evidence. Advances in Insect Physiology, 63, 155-229.

can live up to three years, meaning they may have a longer lifespan in captivity. And because of the care for such a tiny creature, she will be able to live a full life.[93]

How Human Culture Is Shaped by Grasshoppers

In *A Bug's Life*, grasshoppers (the leader cleverly named Hopper) are the bad guys and are made to look extremely intimidating. Unlike the ants in that movie, which the creators tried to make more likable by removing certain insect physicality (four legs instead of six), the grasshoppers, as the villains, were designed to look less human-like. Their size in comparison to ants is also used to make them more intimidating, which is accurate, since they are significantly bigger than most ants.

Sharing many similarities to *A Bug's Life*, "The Ant and the Grasshopper" is an Aesop fable; one among many written by him that includes insects as main characters. In the original, it was actually cicadas and ants but was at one point changed to grasshoppers and ants. The fable states that grasshoppers spent their entire summer singing and dancing, without collecting any food. When winter arrives, the grasshoppers are dying of starvation and beg the ants for help. The ants tell them to "dance the winter away" and do not help. The interpretation of the moral has been different throughout time, but in general, it emphasizes the importance of planning for the future and working hard.

The green grasshopper cocktail became famous in the U.S. in the 1950s. It is green because of the use of creme de menthe and that green made them think of a grasshopper. Now there are grasshopper shakes and cakes and all kinds of treats that carry the name. The only tie it has to grasshoppers is the green color. Mint has nothing to do with them, but it is a famous drink that you can find in many variations in places around the country. I wonder if anyone has made a grasshopper shake with actual grasshoppers in it. It would be less minty but maybe a hit?

In language, the phrase "knee high to a grasshopper" has fallen out of use for the most part but I've heard it a couple of times in rural Missouri. It often accompanies the statement, "Last time I saw you..."

93 Oberholtz, C. (2024, Feb 17). Miniature 'neck' brace helps save grasshopper at Texas zoo. Fox News foxweather.com/earth-space/neck-brace-peruvian-jumping-stick-houston-zoo.

and you can replace grasshoppers with other insects like mosquitoes or bees. "You used to be so tiny but look how you've grown" and then someone pinches your cheek. In Thai, there is an idiom which states "ride an elephant to catch a grasshopper," which captures the idea of investing a lot of time to get something small in return. You may have heard the term "eat like locusts," which is often in reference to a large group of people eating the food provided very quickly, like when the dwarves show up to Bilbo's house in *The Hobbit* and eat everything in his pantry. There are also numerous sports teams named Grasshoppers: Zurich, Greensboro, Sedona, and Columbus are all cities with teams named after these jumpers.

Quick Facts: Grasshoppers

- Every locust is a grasshopper but not every grasshopper is a locust. Not all have the ability to change into a locust.
- An important source of food for all animals including humans. Grasshoppers are a super popular snack at Marlins baseball games.
- They may be able to feel pain and were one of many insects studied to determine the validity of the claim.
- Just because you see a few does not mean they will damage your plants. Rarely are they destructive enough to cause issues in gardens.
- "The Ant and the Grasshopper" was a fable by Aesop that inspired the Disney movie *A Bug's Life*. Grasshoppers are lazy in it, which isn't really accurate to real life (but still a great fable).

Spotlight: Rocky Mountain Locust (*Melanoplus spretus*)

The only ones left are now those that can be found in museum collections.

- This locust was the only one in the U.S. and is now completely extinct.

- The extinction took place over one full year. Entomologists guess that it happened because of habitat destruction (plowing the fields at the end of the year, killing all of their young).

- They had caused a famine in the 1870s which led to the creation of the Entomological Committee in the U.S. (now what would be under jurisdiction of the Department of Agriculture).

- When a grasshopper transforms into a locust, it goes through physical changes, like brighter colors (for visibility) and longer wings (for sustained flight). They gather in larger groups and then have more babies, resulting in a lack of food. This causes them to swarm and decimate crops. Again, most grasshoppers are unable to turn into locust: less than twenty in 11,000 species.

- They were the primary food for a bird that was one of the most common shore birds in North America: the Eskimo curlew. It is believed that the loss of the grasshopper was a factor in the extinction of the curlew. (Not officially extinct, but there has not been a reliable sighting since 1987).

Part VII The Speedy Ones: Dragonflies and Damselflies (Order Odonata)

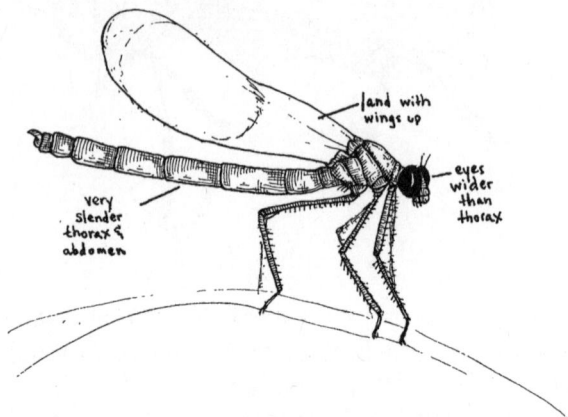

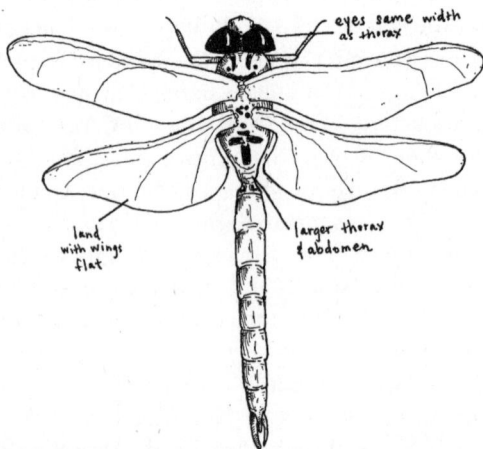

Top: Zygoptera, previously called damselflies. Bottom: Anisoptera.

CHAPTER 18: DRAGONFLIES (INFRAORDER ANISOPTERA) AND DAMSELFLIES (SUBORDER ZYGOPTERA)

*M*any entomologists believe dragonflies and damselflies are the same, even if their outward appearance is different. They use the term dragonfly to describe every species within Odonata. Their similarities would make it impossible to create a full chapter on both. Because of that, unlike every other section, this will only have one chapter. Moving forward, I will refer to both the infraorder Anisoptera and the suborder Zygoptera as dragonflies.

ZYGOPTERA: Head is wider than thorax, land with wings up, generally smaller, more slender thorax and abdomen

ANISOPTERA: land with wings down, often much larger

Dragonflies and damselflies are undeniably cool. If I could be any insect, it would be a dragonfly. Even for people who do not like bugs, watching as these creatures acrobatically fly over the surface of the water is enthralling. They have so much control, feature such bright colors, and are so large. It's easy to marvel at them. They have been around for about 320 million years and have an extremely effective, fine-tuned build. Their scientific name *Odonata* comes from the Greek word for *tooth*, but no one really knows why that was chosen.[94] Because these bugs don't have teeth. Since the designation of that name in 1793, it has stuck. It's most likely going to stick around forever, even though the reason for the original name is lost to time. Their distant ancestors, the griffinflies, were enormous versions of our modern-day dragonflies: wingspans of over two and a half feet (30 in.). The largest one now, the giant darner, only has a five inch wingspan, 1/6th of the size. Insects at that time (about 300 million years ago) were thought

---

94 Mickel, C. E., The significance of the dragonfly name "Odonata", *Annals of the Entomological Society of America*, Volume 27, Issue 3, 1 September 1934, Pages 411–414, doi. org/10.1093/aesa/27.3.411

to be able to get that big because of high oxygen levels. Bugs don't have lungs, so they would need a lot of oxygen to get through their tiny breathing tubes. Some scientists believe that the disappearance of griffinflies coincides with the success and diversification of birds around 230 million years ago [95] meaning that it was because they were food that they disappeared, not tied to difference in oxygen levels.

Regardless, the dragonflies that are still around today are pretty large compared to many other bugs. They are hard to miss and easy to wonder over.

How Dragonflies Shape Our World

Dragonflies in the infraorder Anisoptera are the most efficient hunters in the animal kingdom with a 95 percent success rate. Most mammals have a success rate below 50 percent(although African wild dogs and some porpoises have close to 90 percent). Dragonflies like to attack prey midair. They eat mosquitos, flies, and many other invertebrates, but mosquitos are particularly important to their diet. They are amazing fliers but their eyes play a large part in their hunting success. They have compound eyes that wrap around their head, meaning they can see in almost any direction. Scientists find them difficult to catch sometimes because they can see the net coming from behind them. Many species can also see ultraviolet colors. Humans can see 60 frames per second, whereas dragonflies see 200 frames per second. Dragonflies in suborder Zygoptera are also great hunters but pluck their food from grass blades or branches.

Even with great eyesight, they are tricked by shiny, human creations. Researchers have discovered that dragonflies can mistake the shimmer of polished graves and solar panels for water.[96] When they are confused by these surfaces, female dragonflies have been observed laying eggs. There is no chance of survival for a dragonfly egg laid on a solar panel. Biologists worry that these surfaces, especially with an increased number of them, could cause a significant issue going

95 Stevens, T. (2012, Jun 4). *Reign of the giant insects ended with the evolution of birds*. UC Santa Cruz. news.ucsc.edu/2012/06/giant-insects.html

96 Horváth, G., Malik, P., Kriska, G., & Wildermuth, H. (2007). Ecological traps for dragonflies in a cemetery: the attraction of Sympetrum species (Odonata: Libellulidae) by horizontally polarizing black gravestones. *Freshwater Biology*, 52(9), 1700-1709.

Horváth, G., Blahó, M., Egri, Á., Kriska, G., Seres, I., & Robertson, B. (2010). Reducing the maladaptive attractiveness of solar panels to polarotactic insects. *Conservation Biology*, 24(6), 1644-1653.

forward. Scientists are trying to figure out the best way to reduce the attractiveness of these surfaces to odonates.

Dragonflies go through a metamorphosis only seen in a few other bugs (i.e. mosquitos): their nymphs are aquatic/water reliant, while the adults are terrestrial/land reliant. They do still spend most of their time around water as adults and lay their eggs in water. All stages of dragonflies are important bioindicators of water health. If a pond is unhealthy in some way, you are less likely to see these bugs using the space. Dragonfly nymphs are predators who hunt as effectively as their parents. They are known to eat anything they can get their little toes on underwater, even small fish and frogs. As all things are balanced, they are also a yummy snack for fish that are larger than them. If one of their legs gets nabbed, they can regrow it. They are less lucky when eaten whole. In fly fishing, because of their overall tastiness, many anglers use lures that look like dragonfly nymphs.

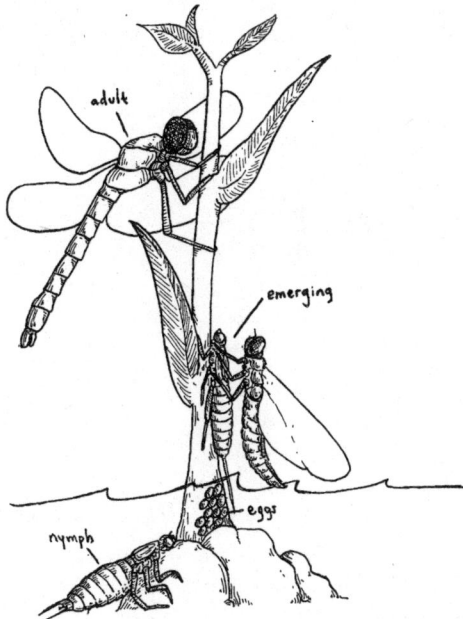

Dragonflies go through complete metamorphosis and shed an exuvia like cicadas. Similar to cicadas, they need to hang out on a plant to let their wings fully form and harden before flying. They are vulnerable during this time.

The mating rituals of odonates are complicated. I'm sure you've seen the action in real life or in pictures. Ever seen two stuck together, flying around? Yeah, they're mating. The male attaches the end of his abdomen to the reproductive organ of the female. They fly around that way for a while, creating what could almost be described as a heart shape. While that may sound romantic, many female dragonflies disagree. Scientists have witnessed females hiding or pretending to be dead to throw a male off of their scent.[97] Females also have sex with tons of dudes, so they can be picky about who they smash. It also seems exhausting because it can last for upwards of multiple hours.

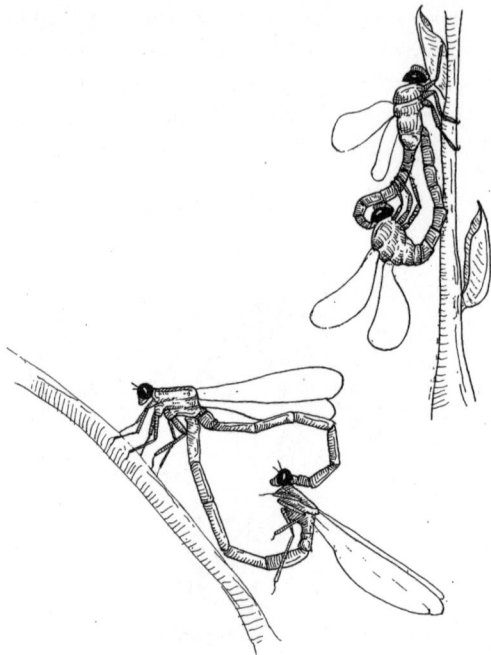

Zygoptera species form a "heart" of sorts when mating, whereas most Anisoptera species mate in a more 69-ish position.

The male's reproductive organ (a very complicated penis) is used not only for injecting sperm into the female but also to scrape out any sperm from a previous mate.[98]

97 Newman, A. (2011, Oct 17). *It's complicated: Dragonfly love comes calling.* New York Times. nytimes.com/2011/10/18/science/18dragonfly.html.

98 Note: Over time, females have evolved ever-complicated reproductive tracts to try and stop that from happening nytimes.com/2011/10/18/science/18dragonfly.html.

How Humans Can Interact with Dragonflies

Some birdwatchers have taken up an interest in dragonflies while they are out looking at birds. Birds eat dragonflies so they have always been around while birders are out. Organizations, like the Audubon Society and the Xerces Society, have collaborated to encourage birders to also track dragonflies. Dragonfly watching isn't necessarily a new activity, but this collaboration could help to keep more accurate population numbers and track their migration patterns. People who exclusively go out to find and observe these amazing creatures call themselves dragon-watchers, which is a pretty hardcore name. Many cities and/or scientific organizations have citizen science projects that ask for help from nature enthusiasts to collect data. Citizen science, as described by NASA, is "collaborations between scientists and interested members of the public" and that "through these collaborations, volunteers (known as citizen scientists) have helped make thousands of important scientific discoveries."[99]

Similar to butterflies, dragonflies are a good representation of charismatic microfauna. Most people like dragonflies, so perhaps if they care about them they will soon start to care about other bugs. Like so many other insects, they can inform us on the impact of the Anthropocene on the ecosystem, like clean waterways. Currently, little is known about the impact on dragonfly species as a whole. Studying insects is difficult for various reasons but if we can get more people observing and keeping tabs on their numbers, hopefully we can learn more.

The Butterfly Pavilion (BP) in Denver is studying the viability of captive breeding programs for endangered and threatened dragonflies in Colorado, specifically the Hudsonian emerald dragonfly. Similar to the Species Survival Plans (SSP) at other zoos, the BP project is at the forefront of understanding breeding dragonflies to be released back into the wild. Many organizations discourage the release of reared into the wild (like was discussed in the butterfly chapter when talking about monarchs) but Species Survival Plans are different. SSP are overseen by the Association of Zoos and Aquariums (AZA). The AZA has extremely high standards for certification and only the "best" zoos are part of the association, as well as participate in SSPs:

---

99 NASA. (n.d). Citizen science. science.nasa.gov/citizen-science/

AZA's rigorous, scientifically based and publicly-available standards examine the zoo or aquarium's entire operation, including animal welfare, veterinary care, conservation, education, guest services, physical facilities, safety, staffing, finance, and governing body. AZA standards are performance-based to allow them to be applied to a variety of different situations and cases. AZA is continuously raising its standards as science continues to learn more and more about the species in our care. Accreditation is rescinded if AZA standards are not maintained.[100]

Only 10 percent of the 2,800 animal exhibitors are AZA Accredited. Zoos with accreditation will always display the AZA logo should they be part of the organization. They are the overarching body for legitimate zoos that can participate in SSP. Butterfly Pavilion is an AZA accredited invertebrate zoo but this project is (not yet) an SSP. They partnered with wildlife biologists to determine how to best rear the nymphs into adulthood and then get the adults to survive and lay eggs. They have been able to get the nymphs into adulthood but have not made it past the next hurdle. The end goal is to be able to reintroduce some of the adults into the wild while relying exclusively on captive adults to lay eggs (they currently collect wild eggs).

In addition to their efforts of breeding dragonflies, they have a Dragonfly Monitoring Program that trains volunteers and then relies on them to collect field data about different dragonflies at different sites. You attend a training (most likely about identification) and then you visit your assigned site, like a park near you, eight times between May and August. You report what species you saw (if you can identify them) and how many you saw. It is a citizen science project that helps keep track of dragonfly numbers throughout the state. Their interest form hasn't been updated since April 2022 however, so not sure if it is ongoing. You can also use iNaturalist without being tied to a specific citizen science project.[101]

---

100 Association of Zoos and Aquariums. (n.d.). About AZA accreditation. aza.org/what-is-accreditation.

101 Note: iNaturalist is a free app and you are not expected to know what you take a picture of. People who can identify the plants/animals/fungi you see will do that for you. All uploads (with clear pictures) are helpful for the scientific community.

How Human Culture Is Shaped by Dragonflies

Dragonflies have individual flight muscles attached to the bottom of each wing, which allows the wings to move independently. This is one of the main reasons they are strong, accurate fliers. Many scientists have tried to create small machines replicating the flight of a dragonfly for accurate, versatile drones and such. They have not yet found success. *Dune*, the novel by Frank Hebert, contains dragonfly planes (but they have three pairs of wings instead of two). Unfortunately, this is still only science fiction. Dragonfly-like aircraft would allow for extremely precise movements and amazing maneuverability. And, more importantly, they would look super dope.

Many cultures have been fascinated with dragonflies. In Diné folklore, they represent "pure water" which ties them to their importance as bioindicators. Japan was once called by another name that translated to "dragonfly island" because in mythology, a dragonfly ate the mosquito that bit the founder of Japan. They are often seen as positive and alluring figures. They also show up often in Japanese literature, in pieces like haikus. Some areas of the world, like Europe, did not see them in a positive light. In English folklore, dragonflies were called the devil's darning needle and were capable of sewing your mouth, nostrils, or eyelids shut should you get on their bad side. The Norwegian word for dragonfly translates to "eye-poker" and one Portuguese word for them translates to "eye-snatcher." Some folk beliefs tie them to snakes: in Welsh, they are called "adder's servant"; in the Southern U.S., they are called "snake doctor." People believed that they collected food to feed the snakes or that they could sew the snakes back up if they were injured (ties into the "devil's darning needle" folk belief as well).[102]

They are also an inspiration for some jewelry and furniture artists. There is an Egyptian amulet from 2500 BC that depicts a dragonfly. Tiffany, famous for their glassware, jewelry, and dishware, used dragonflies as designs on many of their pieces in the early 1900s. Other brands at the height of Art Nouveau designed jewelry giving the people what they wanted: vibrant, large, eye-catching dragonfly pieces. In conjunction with our fascination with them and their abilities, dragonflies have been used in traditional medicine for centuries.

---

102 Hand, W. D. (1973). From idea to word: Folk beliefs and customs underlying folk speech. *American Speech*, 48(1/2), 67-76.

In northeast India, some people use nymphs as blood purifiers, to treat anemia, or to treat upset stomachs. Fighters in Zambia powder themselves with crushed adult dragonflies to help them dodge punches. Japanese traditional medicine uses them for combatting asthma or coughing.[103] Some places also just eat them for nutrition, where they are caught and fried. Eating raw dragonflies (or any insect really) is not encouraged. They are currently not high on the list of people's favorite insects to eat. Some scientists don't believe that dragonflies as farmed food is a plausible idea, meaning they will not likely show up as a species to help food insecure nations on a large scale.

Quick Facts: Odonates (Dragonflies)

- They can bite and have powerful jaws. It rarely happens and only when they feel threatened (like if you completely close your hands around them).
- Dragonflies and damselflies are different: Many entomologists do not believe that to be the case. All infra and suborders in Odonata are dragonflies. The term damselfly is also very rooted in gendered tropes.
- Dragonflies aren't evil: they don't sew your mouth shut and if anything, they are good because of all they do for the environment
- The nymph forms of dragonflies all live in water and are voracious hunters (but also make tasty snacks for fish).
- Dragonflies don't harass horses. They are most probably around horses to eat the flies that pester them. Some people call them "horse stingers," but they don't sting. Their butts are pretty big and often move in a way that looks like they may be stinging. They aren't, and they physically can't.

103 Meyer-Rochow, V. B. (2017). Therapeutic arthropods and other, largely terrestrial, folk-medicinally important invertebrates: a comparative survey and review. *Journal of Ethnobiology and Ethnomedicine*, 13, 1-31.

Bug Spotlight: Common Green Darner (*Anax junius*)

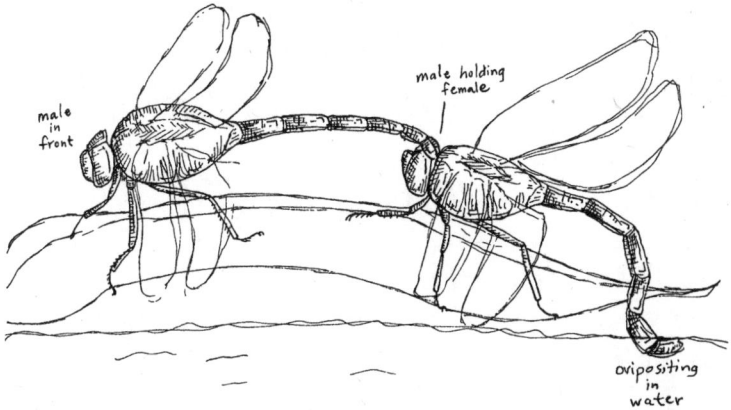

male in front

male holding female

ovipositing in water

Tandem ovipositing

- The dragonfly most often associated with the beginning of summer and one of the most common species in the U.S.
- They can be around three inches long and up to four inches wide, so they are fairly big.
- They migrate from northern Canada to southern Canada and can cover up to 87 miles a day.[104]
- One study found that they have an average flying speed of 10 mph.[105]
- The female lays eggs, which the male holds onto her head, which is uncommon in dragonflies

104 Knight, S. M., Pitman, G. M., Flockhart, D. T. T., & Norris, D. R. (2019). Radio-tracking reveals how wind and temperature influence the pace of daytime insect migration. *Biology letters*, *15*(7), 20190327. doi.org/10.1098/rsbl.2019.0327

105 May, M. L. (2013). A critical overview of progress in studies of migration of dragonflies (Odonata: Anisoptera), with emphasis on North America. Journal of Insect Conservation, 17, 1–15.

Part VIII How Do These Even Go Together?: Praying Mantids, Termites, and Cockroaches (Superorder Dictyoptera)

Praying mantis, termites, and cockroaches—I was genuinely shocked to find out that these three were in the same superorder. I had never even heard of it because I had always seen them referred to by their order names: Mantodea (mantises) and Blattodea (cockroaches and termites). Only within the last ten years have termites (order Blattodea, infraorder Isoptera) been officially included in this superorder. They all belong to the superorder Dictyoptera. While they may not look similar on the outside, scientists know that their internal similarities make them part of the same order. One of the similarities is the ootheca (ō-ə-ˈthē-kə, o thee ka). Besides being the coolest word ever, it is the egg sac that cockroaches and mantises lay that contain their growing babies. Termites may or may not lay oothecae. They may actually represent the ancestral form of what an ootheca became[106]. Meaning that what termites lay is the original form of an egg sac, whereas mantids and cockroaches evolved to have slightly different (and harder) ootheca. Their other inward similarities are way above my head in terms of morphological knowledge/understanding, but trust that other scientists know what they are.

---

106 Nalepa, C. A., & Lenz, M. (2000). The ootheca of Mastotermes darwiniensis Froggatt (Isoptera: Mastotermitidae): homology with cockroach oothecae. *Proceedings. Biological sciences, 267*(1454), 1809–1813. doi.org/10.1098/rspb.2000.1214

American
Cockroach

— actual size

European Mantis

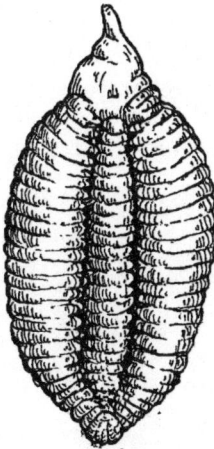

actual size.

Termite eggs

— actual size

Mantid and cockroach oothecae are fairly hard in order to give protection to the larvae growing inside. Termite oothecae are softer but they lay them within their colony so they don't need as much defense.

CHAPTER 19: COCKROACHES (ORDER BLATTODEA)

*T*think it is important to talk about cockroaches first. They are universally disliked and the subject of many horror movies/images. They don't necessarily deserve that bad of a rap. Trust me, I've had some horrendous run-ins with American cockroaches (the REALLY large ones): one crawled over my fingers while I was sleeping, one crawled up my legs while I was sleeping, one flew onto me while I was sleeping in the living room hoping to get away from the roaches in my bed. And we didn't have an infestation. The first time I saw one I thought it was a mouse based on the size. Even my cat Charlie looked at it and was like "Nah, I'm good."

Their suborder name *Blattodea* is from the Latin word *blatta*, which is what they called cockroaches (translates to "insects who shun light'"). Pretty straightforward scientific name. The common name *cockroach* comes from the Spanish word *cucaracha*.

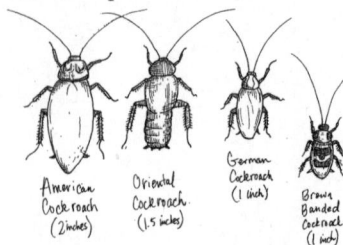

American cockroaches are not a sign of infestation. If you see any of these other species, you may have an infestation.

How Cockroaches Shape Our World

Of the 4,600 cockroach species that occur worldwide, only four of them are considerable pests in the United States: American cockroaches (who are not indicative of an infestation) and the German cockroach, the Oriental cockroach, and the brown-banded cockroach (all three may be indicative of an infestation). That is less than 1 percent. Worldwide, only thirty species total are known to be associated with humans and human dwellings. American cockroaches are interesting because seeing one is not indicative of an infestation. Most of the time, they are passing through on their way to somewhere else, and you often see them alone. They prefer sewers and pipes. After a rain, they can be seen coming out of the sewers just chilling, waiting for the water to subside. They are large (around 2 inches) and very fast but mean no harm.

German cockroaches (*Blattela germanica*) is a worldwide species that is the ultimate roach pest. Seeing one is often indicative of an infestation. Some infestations are bigger and more of an issue than others but seeing them should alert you to a potentially large number in your house. This species has become resistant to most household poisons. Companies are trying to create safer poisons (since in the past, there was little concern about how it affected people) as well as new chemicals to attack the hardy roach. Similar to weevils, it is an ever-evolving battle between humans and bugs. During my first term of AmeriCorps, we had a building which all of the programs used as a home base. Every month, a different team was in charge of cleaning the building, including doing the dishes and disinfecting the bathroom. I was in charge of loading the dishwasher. The moment I opened the dishwasher, I found myself transported to a horror movie: German cockroaches quickly scattered back into the holes of the dishwasher, having been covering every dish seconds before. I pointed out the issue to my boss, and they said "Just be sure to use high heat to kill them all." That practice doesn't actually work. I never used those dishes again (and I discouraged others to use them too). It was a very old building, connected to other old buildings, and often there isn't much you can do. Bombing your place is expensive, time consuming, and uses chemicals that are harmful to humans, making treatment prohibitive for some. It also doesn't guarantee their eradication: they can easily return. Part of their pest designation is because they are so hard to get rid of (like bed bugs or fleas).

Most species can go significant periods of time without water or food: upwards of a year without needing water and up to three months without food. In high school, a teacher cut off a cockroach's head and everyone marveled at how long it was able to live without one. What kills them is lack of food or water after several weeks. They don't bleed to death (because of the way their innards work) and they can breathe through their spiracles (the little holes in their abdomen, like on the sides of caterpillars). They can withstand more radiation than humans (but not as much as fruit flies).

In the *Fallout* games, they appear as radroaches, extra large roaches thanks to dangerous amounts of radiation. Some of them even glow (purely science fiction).

They also live for a long time compared to most insects: some female American cockroaches can live for up to two or three years. Throughout their lives, they can have upwards of 400 offspring. Some females can produce asexually (and the ootheca includes only female roaches) while some species only need to receive sperm once in their lifetime to produce fertilized eggs.

Cockroaches can lay a serious number of eggs, but some of those same species may build loving families: wood roaches have lifelong relationships. Parents take care of their brood for three or more years. One big, happy family. Other species have been found to care for their young for varying amounts of time. Some roaches, most often females of the species, will hang out with their newly hatched young for a few hours before the young head off into the world. Others will allow their young to group beneath them or under their wings, where they then secrete a milk-like substance for the nymphs to feed on. This relationship can last for days to months, depending on the species. Male roaches are often not involved, but some groups consist of a male and female roach that spend time with young roaches of various ages. Scientists don't quite know if the young are all related to the adults, but the care aspects (e.g., protection, feeding) are present. Studies have shown that young can identify their parents and siblings. They believe it is because adults produce distinct pheromones. This level of care is very rare in other arthropods; however, some species of roach also lay their ootheca and walk away, never to interact with their young in a parental way.[107]

While this information may not cause you to love cockroaches, their importance in the ecosystem is undeniable. Without decomposers like cockroaches, we would be completely buried in waste: trash, actual shit, dead things, etc. They love eating our trash and decay, which should be appreciated. We definitely don't want to be the ones responsible for that. Some cultures farm cockroaches (specifically American cockroaches) for traditional medicine (e.g., gastrointestinal issues or

107 Nalepa, C. A., & Bell, W. J. (1997). Postovulation parental investment and parental care in cockroaches. The evolution of social behavior in insects and arachnids, 26-51.

heart failure)[108] and cosmetics (e.g., face creams, face masks).[109] Even in the countries where this practice is common, like China, owners keep the farms a secret because of the stigma. They know that their neighbors don't want to find out they are living next to a roach farm.

How Humans Can Interact with Cockroaches

Most cockroach species are solitary, decomposers hanging out and eating on the forest floor. The ones that are considered human pests have adapted well to urban spaces and want some of the things that humans provide. They love sugar, and who doesn't? A study was done to show that cockroaches love beer (it's actually the sugar they love), so you can share a beer with a cockroach should you choose to do so. In large enough numbers, they can cause allergic reactions in some people. The allergy is rare but it does exist. It is also true that some species in large enough numbers can carry disease and spread illness. They live in sewers and eat decaying things so they naturally are around things that could make humans sick. You may then be surprised to learn that cockroaches are in fact very serious about their grooming. They are constantly cleaning off their feet and antennae (which is actually how most poisons work so well on them. They walk in the poison and then clean their feet, leading to ingestion.)

They weren't considered a pest when they showed up 250 million years ago but times have changed. Humans have changed everything. Roaches are extremely successful in the human manufactured world we've created. Pests are defined based on human parameters. Bethany Brookshire, a science journalist and author, penned the book *Pests* (2022), which states "in theory, we wanted these animals to live. We just didn't want them to live here." And that when they do, "It's the people who get the blame. Animals become a symbol of the 'sins' of the people living with them," even extending into what we call others.

Scarface, Al Pacino's character in the 1983 film, uses the term *cockroach* to describe many of the people that he is unhappy with or doesn't trust. It is a commonly used name for a person(s) who are "undesirable" or "procreating rapidly."

108 Zeng, C., et al. (2019). The role of Periplaneta americana (Blattodea: Blattidae) in modern versus traditional Chinese medicine. Journal of Medical Entomology, Volume 56, Issue 6, Pages 1522–1526, doi.org/10.1093/jme/tjz081.

109 Ukoroije, R. B., & Bawo, D. S. (2020). Cockroach (Periplaneta americana) Nutritional value as food and feed for man and livestock. Asian Food Sci. J, 15(2), 37–46.

How Human Culture Is Shaped by Cockroaches

Popular culture sometimes adds to the stigma and fear around cockroaches. One of my favorite *X-Files* episodes, "War of the Coprophages," involves deadly cockroaches. Coprophages are animals that eat poop. They eat waste, which includes poop, but they do not eat that exclusively (dung beetle larvae are coprophages). Everything about the episode makes cockroaches repulsive: they cause anaphylactic shock and four deaths, they swarm, there are walls visually crawling with thousands of roaches. The best part is the entomologist in the episode who studies them: an attractive, young woman named Bambi. She was an inspiration to me as a kid. It was supposed to be funny and surprising that someone so attractive was interested in something so unattractive.

Some people tolerate or enjoy cockroaches for varying reasons. They are used in races at Insect Expos. Humans will bet on anything. Many species that are not considered pests are kept as pets or as food for other pets. Dubia cockroaches are frequently kept, bred, and given as fun snacks to lizards. Madagascar hissing cockroaches and death's-head cockroaches are kept as pets. These two species are also often used as educational animals at zoos, allowing children (and interested adults) to hold them. While volunteering at the Butterfly Pavilion, I was in charge of the hissing cockroach display and allowed people to "pet" it with one finger. The talking points were how many roaches are not considered pests and they mostly ate detritus from the forest floor. People were always surprised to learn that and softened towards them. They didn't seem like they planned to embrace roaches any time soon but had a positive fact to share about them.

Some digital media has tried to make them more relatable and likable; *Wall-E*'s best friend is a cockroach and the animators did a great job of making the roach accurate while also making it cute. *Joe's Apartment*, a 1996 movie, has talking and singing cockroaches. The movie was panned, but it gave roaches a more upbeat, human-like quality while also keeping their physiology accurate. I can't say I'd recommend the movie, but I like what they did with the roaches. The song "La Cucaracha" is about a roach who lost one of his six legs and finds it difficult to get around on his remaining five legs. I used to think that the song was about trying to step on a cockroach, but the actual lyrics are more sympathetic.

Quick Facts: Cockroaches

- Most don't live in human dwellings even if you see them (like American cockroaches).
- Getting rid of a true infestation is difficult so seek the help of professionals.
- Cockroaches are very clean. They frequently groom their antennae and feet.
- They do not change much when exposed to high levels of radiation, leading to the common quote in post-apocalypse media about how all that is left are cockroaches and twinkies.
- Can go for long periods of time without food and water. When you cut their head off, they can stay alive until they die of hunger or thirst. This is because their brain is located a little future back in their thorax than other insects.

Bug Spotlight: Palmetto Bugs

A. Florida woodroach, *Eurycotis floridana*. B. Smokey brown cockroach, *Periplaneta fuliginosa*. C. American cockroach, *Periplaneta americana*.

In the South, there is some confusion over the differences between a cockroach and a palmetto bug. They are actually the same thing. Palmetto bugs refer to where the bugs are often seen, which is palmetto trees. In South Carolina and Florida, there are a few species that may be referred to as Palmetto bugs: the American cockroach, the smoky brown cockroach, and the Florida woods Cockroach.[110] None of these species choose to live in human dwellings and are more likely to be seen outdoors (perhaps hanging out on a palmetto tree). So, palmetto bugs are cockroaches, but the name can possibly be attributed to a few species. Calling one a palmetto bug is totally okay, but when someone argues with you that it is just a roach, know that they may be correct too.

---

110 Mundorf, D. (2023, Jan 30). Palmetto Bug vs Cockroach: What's the difference? *Bob Vila*. bobvila.com/articles/palmetto-bug-vs-cockroach/

Chapter 20: Mantises (Order Mantodea)

*T*hate to admit it but oftentimes my Freds (crickets I captured) would be given to my Mannys (mantises that I captured). (NOTE: I am very aware that Manny is a wholly uncreative name for a mantis but I also had a lizard named Lizzy. I swear I'm more creative now.) If you've ever seen a mantis eat, it is a sight to behold. They immediately go for the brain to stop their prey from fighting back. They do the job quickly and can even take on insects the same size as them (and sometimes even the same species as them). Mantises really fit the alien description that is often attributed to bugs: the shape of the face, the green color, the long slender body. They are masters of camouflage, making them difficult to spot. They are also fairly large and uniquely built. Since they are rarely found in houses, they are less likely to be considered a pest than their superorder-mates, roaches and termites.

Mantids, in the suborder Mantodea (prophet-like in Latin), are fairly "new" on the insect evolutionary tree. Few fossil records exist of mantises, but they first show up around 145 million years ago, long after roaches. Praying mantises get their name from the prayer-like position of their arms. The common European mantis's scientific name is *Mantis religiosa*. Worldwide there are 2,400 known species, many of whom are spectacular looking (check out the orchid mantis or the dragon mantis). More continue to be found. Thanks to some citizen scientists on iNaturalist, two new species were classified in Australia in December 2023.

How Mantises Shape Our World

In the U.S., two of the most abundant species are non-native and were accidentally introduced from Europe and Asia, but are now here to stay: the Chinese mantis and the European mantis. Once they were discovered to be in the U.S., they were bred to produce larger numbers for pest control, but there are still no studies that show they are beneficial to garden plants.[111] Little is known of their effect on

---

111 Klass, K.D. & Grossmann, C. (2009). Mantophasmatodea. In Resh, V. H., & Cardé, R. T. (eds.), *Encyclopedia of Insects.* (2nd edition, pp. 599-600). Elsevier.

native species. The Carolina mantis and Arizona mantis are native but mantis populations are difficult to study, which means their population numbers may not be accurate. There are few studies of species/suborders as a whole, especially in regard to anthropogenic effects (issues caused by humans).

In the summer of 2023, scientists were excited to find people posting photos online of mantises eating the invasive spotted lanternfly. The University of Pennsylvania alone has received over 600 photos of mantises eating lanternflies over the past few years.[112] And those are just the ones caught on film. It seems that mantids and chickens really like to eat them, so people are hopeful that populations will diminish with more native animals feeding on them. In their native China, there is a parasitic wasp that keeps numbers down, but perhaps in the U.S., the mantises (even the non-native ones) can help control lanternflies.

Morphologically, mantids are pretty unique. They have what are called "raptorial" front legs, which are useful when they hunt. The legs are lined with small spikes that help to hold their prey. Their diet mostly consists of arthropods, but they have been known to catch and eat lizards, fish, and small birds. They have very large eyes that can see well during the day and at night but are most active during the day. If you look at a mantis, they appear to have a pupil. It is in fact what is called a pseudopupil. This fake pupil does not function like a human pupil, but it does give their eyes a more human-like appearance in comparison to other insects. Researchers report seeing males move frequently at night in search of a mate. Birds are their biggest predator during the day so traveling at night is safer. Similar to katydids, they can sense the echolocation of bats so they don't move if bats are close.

If they are threatened, mantids display anti-predator strategies. First, they stand up tall and put their raptorial legs out to the sides while flaring out their wings. Most species have spots and colorful patterns that are meant to intimidate a predator. Some species are even capable of creating a hissing sound with their abdomen. They will strike with their forelegs if a predator gets too close. One of their more interesting predators is a parasitic worm called a horsehair worm. These worms reproduce in the water and then the small, baby worms move into an aquatic/semi-aquatic insect. When the mantis eats that

112 Wesser, J. (2023, Aug 2). This insect is the spotted lanternfly's worst enemy. *ABC news.* abc27.com/digital-originals/this-bug-is-the-spotted-lanternflys-worst-enemy/.

insect, it lives and grows inside the mantis. In order to complete their life-cycle, they kind of take over the mantis and lead it to water where it then drowns itself. The worm itself doesn't kill the mantis, but the worm drives the mantis to kill itself.[113] Another video to look up only if you have a strong stomach: the worm then leaves the carcass and lays its eggs. Some people who capture adult mantises will place their abdomens in water to expel any present horsehair worms.

A Carolina mantis, *Stagmomantis carolina*, in defense position. Notice the eyespots on both its arms and wings.

A common myth about mantids is that the females eat the males after sex. This behavior has most often been observed in laboratory settings when the males have nowhere else to go after copulation. Scientists believe that it doesn't happen often in nature. When it does, it happens close to the end of the season when the female is about to lay her ootheca and needs serious nutrients when all other food is becoming scarce. Anecdotally, some males in certain species offer themselves to the female. There is a possibility that she will eat the male but it isn't a guarantee. The presentation of praying mantises as "femme fatales" is over-hyped. Once mating is over, the females lay an ootheca, similar to a roach. Unlike cockroach ootheca, mantis ootheca are textured, almost taking on a scrotum-esque appearance.

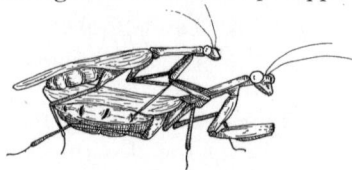

The male is the most vulnerable during mating. One, because they are on the back of the female, meaning they will be attacked first by a predator (serving as a shield for the female). Two, if it is close to the end of the season, they are more likely to be eaten by the female afterwards.

113 Incorvaia, D. (2024, Jan 1). How a parasitic worm forces praying mantises to drown themselves. *Scientific American*. scientificamerican.com/article/how-a-parasitic-worm-forces-praying-mantises-to-drown-themselves/

Camouflage plays an important part not only in their survival but in their hunting success. If you've ever seen a mantis walk, they look like they are trying to avoid sandworms in the dunes of Arrakis: offbeat, swaying steps that look like they might be having trouble moving properly. Some scientists believe this movement is meant to mimic leaves and grass blowing in the wind, which is what they best blend in with; however, some think it may help them discern items in front of them. Many mantids are brown or green in color, but some specialized species like the orchid mantis are meant to blend in with specific flowers (like orchids) where they wait for prey. This camouflage keeps them secret to predators and prey. Some species even mimic other insects. The recently hatched mantises look like ants so they are less likely to be preyed on by ants.

From what we do know, scientists worry that global warming is having a devastating effect on mantid populations. They worry warming will force them poleward too quickly, and potentially cause regional or perhaps even global extinction. Scientists find it difficult to determine mantids' ecological role in ecosystems since they feed on many trophic levels (cannibalism, pollinators, predators, pollen), and it is undetermined how much they affect the niche they inhabit.

How Humans Can Interact with Mantises

Stories of them in fir trees have become more prominent over the last few years. Apparently, females will lay their egg sac (also called *ootheca*) in trees and some northeastern species seem to prefer pine trees. People will go out to cut down their own Christmas tree and then find hundreds of mantis babies weeks later. An acquaintance of mine from New York was doing yoga in their living room a week or so after bringing their hand-cut tree into their apartment. While in downward dog, they saw a small baby mantis. They thought it was cute, said hello and curiously asked how they got into their twentieth story apartment. Very quickly, they realized that their entire floor and tree were covered in dozens upon dozens of baby mantises. For them, it quickly went from cute to scary. Thankfully, baby mantises are not harmful to humans, but seeing that many in your home unexpectedly can be quite the shock. My recommendation would be to not panic and put as many of them into a jar to be released outside. If that idea terrifies you, please don't tell me what your actions would be.

I caught mantises from the wild as a kid, but the captive breeding mantis pet trade is booming. They are one of the most commonly kept insects. Currently, there are basically zero regulations around the mantis trade. Researchers surveyed over 100 mantis owners and they found that many of the participants have no idea of the laws surrounding sending animals across country borders.[114] The majority of packages that the live mantises arrive in does not mark them as such to avoid customs. A surprising number of participants caught wild mantises or purchased ootheca from native species like the Carolina mantis. They mention that wild caught specimens may cause problems with local communities, especially if large amounts of oothecae are being gathered to sell. They also worry about non-native species being released, though they believe introduction of non-native species is more likely to happen during commercial trading, not the pet trade. Most people who buy/find mantises are good at keeping them throughout their entire life cycle and not releasing them. Battison calls for more regulations of the trade as well as more research on how wild caught/collected specimens affect native populations.

Some science educators believe that mantids can serve as a great flagship species.[115] Flagship species are animals that help garner interest in other animals in that group or geographical region. Connecticut's state insect is the European mantis; however, students are trying to remove it as the official insect. Instead, they want to replace it with a native species. They feel that a native insect could help encourage people to care about local environmental issues. The potential replacements unfortunately do not include a native mantis but instead a dragonfly and a butterfly. South Carolina has the Carolina mantis as its state insect. Overall, these bugs are interesting to people but little is known about them. Creating interest and wonder around mantids, such as by elevating them as state insects, may lead people to be more interested in bugs in general.

114 Battiston, R., Di Pietro, W., & Anderson, K. (2022). The pet mantis market. *Journal of Orthoptera Research*, 31(1), 63–68.

115 Snaddon, J. L., & Turner, E. C. (2007). A child's eye view of the insect world: perceptions of insect diversity. *Environmental Conservation*, 34(1), 33–35.

Schlegel, J., Breuer, G., & Rupf, R. (2015). Local insects as flagship species to promote nature conservation? A survey among primary school children on their attitudes toward invertebrates. *Anthrozoös*, 28(2), 229–245.

How Human Culture Is Shaped by Mantids

Culturally, mantises are commonly used to invoke horror, usually appearing much larger than they are in real life. They are a staple in insect fear films (like the one described in chapter 4), with films like *The Deadly Mantis* (1957), about a mantis that threatens all of humankind (kaiju-esque). There is an actual mantis kaiju in *Son of Godzilla* (1967). The cover for R.L. Stine's book *A Shocker on Shock Street* has a giant mantis looming over buildings and cars, which then shows up in the *Goosebumps* movie (2015). Overgrown mantises try to eat you in the video game *Fallout New Vegas*, but you can also eat them if you are able to kill them first. If mantises are not horrifically oversized, they are presented as a femme fatale. In the first season of *Buffy*, a mantis woman tries to mate with and then eat Xander. Not only is that inaccurate, it's also extremely problematic since Xander is like 16.

While they may often be portrayed as scary, they also have been thought of as wise and even supernatural by many cultures. Two southern African tribes revered the mantis. In the Egyptian Book of the Dead, a mantis (called Bird-Fly) was responsible for helping the dead reach the underworld.[116] In the movie *A Bug's Life* (1998), the mantis is the magician in the group. He is presented as refined and intelligent, with a British accent (perhaps a reference to the European mantis). In the *Kung Fu Panda* franchise, Mantis is a recurring character who works closely with the main character, Po. In fact, two martial arts stances are based on the mantis: one called The Northern Praying Mantis, the other The Southern Praying Mantis, based on the parts of China in which they originated.

Quick Facts: Mantids

- The two most abundant species in the U.S. are introduced species: European and Chinese.
- It is not common for mantid females to eat mantid males after sex. It is more likely to happen towards the end of the season when less food is available, as well as when in captivity.

116 Prete, F. R., Wells, H., Wells, P. H., & Hurd, L. E. (1999). The predatory behavior of mantids: historical attitudes and contemporary questions. *The praying mantids*, 3–18.

- Mantises as pets are a touchy subject. Entomologists suggest not taking ootheca from the wild and not keeping non-native species as pets.
- They can eat things the same size as them. They start by eating the brain first. Rarely, they have been spotted eating birds.
- They are commonly used in insect fear films or in media to present them as scary. My favorite version is in *Fallout New Vegas* and you can make mantid armor. So cool.

Bug Spotlight: Carolina Mantis (*Stagmomantis carolina*)

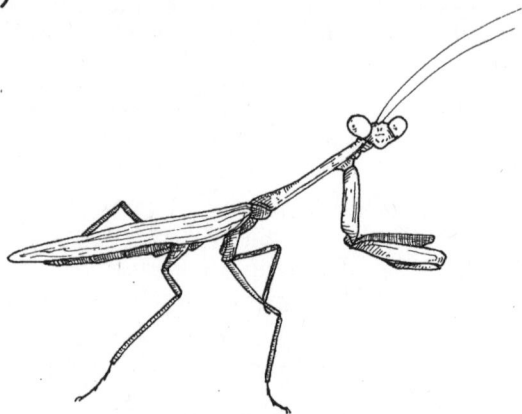

What a freaking cutie.

- The only well known native species of the U.S.
- It is the state insect of North Carolina.
- It kind be found all the way south to Brazil.
- Great at pest control and is the most abundantly seen species in the U.S.
- Females are bigger than males. Both can fly, but males are better fliers overall.

CHAPTER 21: TERMITES (SUBORDER ISOPTERA)

*T*n 2009, termites were in their own order: Isoptera, meaning "equal winged." but since the publishing of the *Encyclopedia of Insects*, they have been placed in the same superorder with mantises and cockroaches. Scientists have been debating this for quite some time, especially after noticing internal similarities between wood roaches and termites. What a strange collection of insects that belong to the same order. Rarely would a person confuse a roach or a mantis with a termite. Outwardly, they all look completely different. Most people confuse ants and termites, who are only in a distantly related order (Hymenoptera). They are sometimes even called "white ants." There are a few ways that you can tell if it is a flying ant or a flying termite, same for if it's a wingless ant or wingless termite.

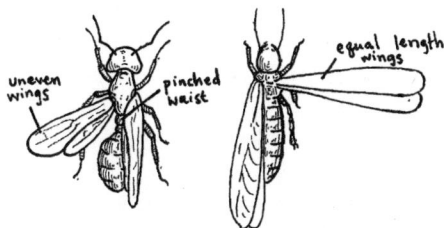

Apart from differences in body shape, they differ in color as well. Ants are often darker brown or black, whereas termites are lighter brown or cream colored.

Fossils of termites have been dated to about 150 million years ago. In 2024, a male and female termite were found preserved in amber from 34 million years ago. The study leader, Nobuaki Mizumoto from University of Alabama, was so excited to find it, not only because it included a male and female but also showed a behavior: mating. Fossils or amber don't often show a behavior, so scientists call these types of finds a "unicorn," since they are so rare. The preservation shows that their mating behavior has been the same over the last 34 million years.[117] They are highly social, similar to ants, and have very complex

117 Langley, L. (2024, Mar 11). Termite fossils caught in the act prove mating hasn't changed in 38 million years. *National Geographic.* nationalgeographic.com/animals/article/amber-fossils-termites-ancient-mating

societal structures. They were the first species on earth, not just insects, to create a complex social system. The field of sociobiology, popularized by E. O. Wilson, frequently uses termites as examples of social behavior. (Note: Sociobiology is an interesting but not well-regarded field that has been debated constantly by biologists since the 1940s. Read at your own risk and know that many scientists do not agree with the research. Many fear it can be too easily tied to eugenics.)

How Termites Shape Our World

People often think of termites as pests since we are most likely to encounter them when they are eating a house. Of the fifty species that live in the U.S., only five are encountered in homes (more on them later). In the wild (which is where most live), they are extremely important to soil health. Similar to earthworms, they are critical for the production, turnover, and enrichment of the soil. They also help to aerate the soil and increase water flow, hence reducing water runoff and soil erosion. They are critical recyclers of wood waste and many areas in which they are native rely on their assistance in creating healthy soil.

A colony is founded by a king and queen (unlike ants and bees who only have queens). When they are looking for a spot to start a new colony, they have wings to fly. Once they pick their location, they lose their wings. Termites eat wood, so they must settle in or close to wood. The couple makes a little chamber for themselves and gets the process started. Queens can be huge, about the size of a human thumb, and can even extend their abdomen to produce more eggs if needed. She can lay upwards of 1,000 eggs per day. Once they hatch, the king and queen take on the responsibility of feeding the young. When the young are old enough to feed themselves, they begin work on the colony and then take over for feeding the king and queen. At that point, the king and queen no longer feed on wood and are instead fed a special royal diet that brews in the guts of their workers. Queens can live upwards of twenty years but some Aboriginal people in Australia claim that the termite queens there can live for 200 years. There is currently no way to date the age of queens.

Workers and soldiers, who make up the rest of the colony, only live for a year or so. Also in a departure from ants and bees, the workers and soldiers are both male and female. They have distinct morphological differences depending on the job they have in the nest.

What is particularly bonkers is that they can even change their body depending on the job that needs to be done. If they are a worker but more soldiers are needed, their outward appearance changes to be like a soldier. They also all look extremely different:

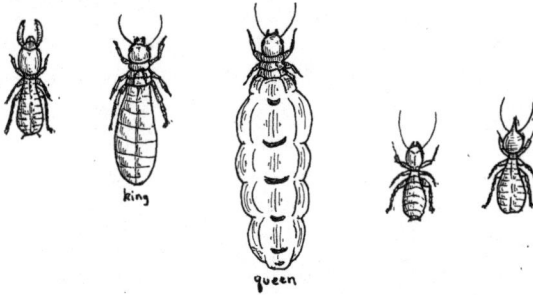

Each caste has very specific positions. You are most likely to see the first termite and the last two: mandibular soldier, worker, nasute soldier. The king and queen are deep in the colony.

Their colony numbers can grow into the millions. While termites and ants comprise about 1 percent of animal species, they represent more than 50 percent of insect biomass since colonies can be so huge.[118] They create and defend while also being blind. Communication between them is extremely important since they cannot see what is going on. Different modes are used to communicate like pheromones, clicking the segments of their abdomen, or headbutting soldiers. This tells the soldiers of imminent danger, so be ready. Workers can also secrete a glue-like substance (sometimes toxic) to defend against invaders. This glue may also include pheromones that alert soldiers to the attack. The glue causes smaller intruders, like ants, to get stuck to buy time for the soldiers to appear. Soldiers cannot feed themselves because their jaws are too large to eat wood or cellulose. They rely on worder termites to feed them. Their poop is an interesting shape and is also the first sign people see of termites in their home. They push the poop out of their nest when there is no longer space for it, leading to piles of it on the ground.

Australia really seems to be the place to study termites. They have magnetic termites but they also have pollinating termites. A critically endangered orchid, the western underground orchid, grows in Western Australia only. The blooms are frequented by termites and

118 Eggleton, P. (2020). The state of the world's insects. *Annual Review of Environment and Resources, 45,* 61-82.

ants who may help with pollination. They travel into the flower, get pollen, and then visit another. It is the only species of flora that is tied to termite pollination.

Other animals also enjoy their mounds as they are well-made and offer ample protection. Ants have been known to move in once termites move out, or even fight them for the mound (could result in a win or a loss depending). Some animals even live in the mound when the termites are still there. Headlight beetles, a bioluminescent coleopteran, live in termite mounds. In there, they are safe and offer a sweet sticky substance that the termites enjoy. When they become adults, they leave the mound to search for another. The relationship is mutually beneficial.

Lacewings, who also live in mounds, are a different story. They are hunters who feed on termites but do so sneakily. The larvae live in small cracks in the mound. When termites come to fix the crack, the larvae kill the termite and continue on like that whenever one comes to check on the crack.

How Humans Can Interact with Termites

Trust me when I say, termites are definitely not trying to mess up your house. They are just eating and it happens to be your house. Obviously, that can be devastating to the homeowner and the structure, which is no joke. As mentioned above, five species are the most commonly found termites in houses: Subterranean (need ground), Formosan termites (ground, introduced from China), Dampwood termites (need ground), Dry termites (don't need to be touching the ground) Conehead termites (introduced species from Caribbean). Four of those five are only attracted to damp wood, and one is attracted to dry wood. The four that are attracted to damp wood will most likely only move into your home if there is a leak or high humidity creates moldy environments. They will only build a colony underground, so it would only be a problem if the damp wood from the house is touching the earth. Dry termites will move into your home only for dry wood. They are often found in the arid southwest and are common in cities like Phoenix, Arizona.

They are also more likely to be found in older homes. Many new homes are built with termite-treated wood. Obviously, this doesn't guarantee a lifetime free of termites, but it greatly reduces the odds.

People in the U.S. spend about $1 billion per year on termite treatment. Many of the treatments are not only detrimental to termites but can kill other insects in your yard. The EPA has a list of things to do to reduce your odds of ever getting them:

- Check for leaks and make sure you have proper soil drainage
- Clean gutters and pipes
- Fill in cracks and crevices, especially around the foundation and around pipes
- Don't stack firewood against your house
- Protect your foundation: choose a concrete foundation over wood.
- Use alternative mulch like gravel instead
- Get regular inspections from a pest company

(Note: My two cents are that the suggestion of using alternative mulch seems unnecessary, and I'm convinced that pest companies will tell you to fear termites even if there is little chance of you getting them. They make a lot of money—$11.5 billion in 2022—by instilling fear. I also just don't like pest companies so be sure to take that into consideration when reading my views.) There is no need to live in fear, as long as you are taking necessary precautions. Even if you do get them, you can get rid of them but will need to be vigilant to make sure they don't return.

How Human Culture Is Shaped by Termites

Termites, their behaviors, and their architecture can be inspiring for humans. Many species build solar chimneys in their nests to regulate temperature inside throughout the day. Humans have used these previously but architects are wondering if these chimneys could be useful in modern day buildings to assist with keeping houses/buildings cooler due to temperature increases worldwide. In Africa, some architects have already begun incorporating this design into new buildings as a natural way to regulate temperature. They are also an inspiration for alternative fuels. Termites can create a serious amount of hydrogen from very little food. Researchers are trying to determine how they can turn plant cellulose into ethanol. This discovery could completely change the game of alternative fuels.

A species of termites in Australia can sense the earth's magnetic field and uses that to create their nests. Their nests are quite the sight. The mounds are tall and thin. The edges are lined from north to south, and the broad side of the mound faces east/west. They do this to minimize the amount of sun that hits the nest, allowing them to keep a comfortable temperature inside. There is also some evidence that these termites can sense the earth's magnetic field, which helps them construct their way-finding mounds. Perhaps not a bad idea for human structure either.[119]

Humans also eat termites and have done so for centuries. Termites are a very popular food in sub-Saharan Africa but numbers of colonies are dwindling due to urbanization and the construction of suburbs. They are also used frequently in traditional medicines. In Nigeria, they are used to treat pregnancy sickness. Termites are seen as soothsayers, which is interestingly often attributed to their relative the mantis. They are also used in Brazil to treat asthma, cough, and the flu. de Figueirêdo and their study partners created a table in 2015 that outlines the different uses for termites around the world as food for humans, feed for animals, and as traditional medicine.[120] Termites are often used metaphorically to show what people believe are good qualities in humans: industriousness, cooperation, unity. In the neotropics, termites are viewed as harbingers of death, whereas in Venezuela, a native group views them as guardians. Termites, along with a paper wasp and two stingless bees, guard the four cardinal points according to their folklore.

Quick Facts: Termites

- Even though they look like ants, they are not related. Termites do not have pinched waists and have equal length wings.
- Many termites only become problems in homes when there is a leak and the foundation or wood in the structure is wet and touching the ground.
- They were the first animals, not just insects, to have complex social systems, including different types of workers, soldiers, and leaders (king and queen).
- Similar to earthworms, termites are important for soil health and aeration.

119 Klotz, J. & Jander, R.(2009). Magnetic Sense. In Resh, V. H., & Cardé, R. T. (eds.), Encyclopedia of Insects. (2nd edition, pp. 592-594). Elsevier.

120 de Figueirêdo, R. E., Vasconcellos, A., Policarpo, I. S., & Alves, R. R. (2015). Edible and medicinal termites: a global overview. Journal of ethnobiology and ethnomedicine, 11, 29. doi.org/10.1186/s13002-015-0016-4.

Bug Spotlight: Drywood Termites

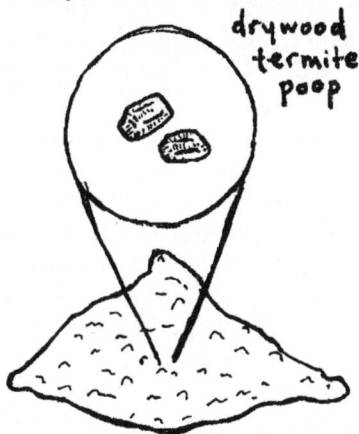

drywood termite poop

It's hard to tell it is poop since it doesn't look like poop from any other species. Very unique in its appearance and texture.

- Unlike almost every other species, these termites don't need to make colonies that touch the ground.
- They are more likely to be a problem in arid conditions, like the American Southwest.
- Less than 10 percent are considered pests or are even likely to show up in your house.
- Drywood termites are the ones with an interesting poop shape: hexagonal. You may first notice little piles on the floor that they push out of a tiny "kick hole" when they need to get the poop out of the colony.
- Similar to bed bugs, companies are beginning to use heat in buildings to rid structures of them. Keeping the temp at 120 or higher for 33 minutes is required.

THE STATE OF INSECTS AND WHAT YOU CAN DO TO HELP

We need to think about and understand how the human world is affecting all bugs—vital parts of our community. Our world has lost 5 percent to 10 percent of all insect species in the last 150 years— or between 250,000 and 500,000 species.[121] The term "insect apocalypse" is a bit of an exaggeration, but we should be concerned. Many insect species are reducing in numbers because of human activity: habitat destruction, pesticides, climate change, introduced species, and pollution. People talk about the windscreen phenomenon. Anecdotally, people would say how they noticed fewer dead bugs on their windshield or grill while driving. Now, scientists have amassed a large amount of research to prove that insect numbers are on the decline, sometimes drastically, essentially validating the anecdotal windshield phenomenon. The Rothamsted Insect Survey, done in the UK, showed that "between 1970 and 2002, the insect biomass caught in the traps declined by over two-thirds in southern Scotland, but remained stable in England."[122] At least numbers remained stable in England.

Many of the surveys have been done in Europe, where very few biodiversity hotspots exist. Even in those areas, decline in populations has been reported. In 2014, scientists did a review of insect related articles in the journal *Science*. For all of the IUCN-documented population trends, which includes 203 insects, 33 percent are in decline.[123] The Krefeld study conducted in Germany claimed that between 1989–2016, there had been a "seasonal decline of 76 percent, and mid-summer decline of 82 percent, in flying insect biomass over the 27 years of study." The decline was "apparent regardless of habitat type."[124] This led to Germany establishing an Insect Protections

121 Janicki, J., et al. (2022, Dec 6). The collapse of insects. Reuters Graphics. reuters.com/graphics/GLOBAL-ENVIRONMENT/INSECT-APOCALYPSE/egpbykdxjvq/.

122 Rothamsted Research Insect Survey. (n.d.). About. insectsurvey.com/about.

123 Dirzo, R., Young, H. S., Galetti, M., Ceballos, G., Isaac, N. J., & Collen, B. (2014). Defaunation in the Anthropocene. Science, 345(6195), 401-406.

124 Hallmann, C. A., Sorg, M., Jongejans, E., Siepel, H., Hofland, N., Schwan, H., ... & De Kroon, H. (2017). More than 75 percent decline over 27 years in total flying insect biomass in protected areas. PloS one, 12(10), e0185809

country-wide program. In a non-European study, Puerto Rico had similar findings. They reported that "biomass losses between 98 percent and 78 percent for ground-foraging and canopy-dwelling arthropods over a 36-year period, with respective annual losses between 2.7 percent and 2.2 percent."[125] Another European study in the Netherlands in 2019 showed an 80 percent decline of butterflies in the Netherlands. The decline was attributed to changes in land use due to more efficient farming methods, which has caused a decline in weeds. The recent uptick in some populations documented in the study was attributed to conservationist changes in land management and thus an increase in suitable habitat.[126] So what humans are doing to combat these issues can help.

Researchers, who did a literature review of all articles on insect numbers, disclosed that in various species in various places, many are on the decline.[127] A conservative number would be an overall loss of 10 percent of insect species.[128] Some entomologists have been calling for deeper investigation into hyper-diverse and understudied taxa to get a better idea of how many insect species we're losing.[129] Biologists want to stress the importance of endangered insects: Scott Hoffman Black and Mace Vaughn, both ecologists who are part of The Xerces Society for Invertebrate Conservation, wrote about endangered insects for the *Encyclopedia of Insects*. In that text, they state that some species play a linchpin role in small, specialized systems, such as caves, oceanic islands, or some pollinator-plant relationships. To conserve insects successfully, the general public, scientists, land managers, and conservationists need to understand the extraordinary value that these organisms provide.

125 Sánchez-Bayo, F., & Wyckhuys, K. A. (2019). Worldwide decline of the entomofauna: A review of its drivers. Biological conservation, 232, 8-27.

126 van Strien, A. J., van Swaay, C. A., van Strien-van Liempt, W. T., Poot, M. J., & WallisDeVries, M. F. (2019). Over a century of data reveal more than 80 percent decline in butterflies in the Netherlands. Biological Conservation, 234, 116-122.

127 Thomas, C., Jones, T. H., & Hartley, S. E. (2019). "Insectageddon": A call for more robust data and rigorous analyses. Global change biology.

128 Watson, R., Baste, I., Larigauderie, A., Leadley, P., Pascual, U., Baptiste, B., ... & Mooney, H. (2019). Summary for policymakers of the global assessment report on biodiversity and ecosystem services of the Intergovernmental Science-Policy Platform on Biodiversity and Ecosystem Services. IPBES Secretariat: Bonn, Germany, 22-47.

129 Isbell, F., Balvanera, P., Mori, A. S., He, J. S., Bullock, J. M., Regmi, G. R., ... & Palmer, M. S. (2023). Expert perspectives on global biodiversity loss and its drivers and impacts on people. Frontiers in Ecology and the Environment, 21(2), 94–103.

They go on to state that people may not have much love for these animals but perhaps people can be swayed based on their importance to human well-being.

The main recommendations for addressing these declines is to counteract habitat loss, reduce the number of pesticides being used, and address ways to slow climate change.[130] In order to do that, we need to protect habitats, change laws/legislation, and increase education about bugs. Some steps can be taken at your home to promote healthy, bug-friendly spaces. I like to call it the "I'm not lazy, I'm just saving the world" approach.

- *Don't cut the grass too often*: insects like long grass to hide in and create their nests
- *Don't pull your weeds if they are flowering*: scientists are torn on this, but pollinators love dandelions and other flowering weeds. If you can't afford to plant anything, it's better than nothing.
- *Don't pull down your dead plants or rake your leaves*: Many insects, including solitary bees and moths, will overwinter in dead plants (also dead leaves). Be sure to leave it all there so that they can keep sleeping.
- *Keep your lights off at night*: many insects can be disoriented by light so keeping them off helps them locate what they actually need to see.
- *Plant native pollinator flowers (if you can)*: This one isn't actually easy since you need space and money, as well as knowledge of what to buy. If you do have space, time, and money, look at the Xerces website to determine what is native in your area. Many cities give away free seed packets. Be sure to ask if they are free of neonicotinoids.

To keep them out of your apartment/home without the use of pesticides:

- *Check and fix any leak*: As was discussed in many chapters, a lot of "pests" like humid, wet, moldy conditions, so be sure to find and fix any leak early.

---

130 Forister, M. L., Pelton, E. M., & Black, S. H. (2019). Declines in insect abundance and diversity: We know enough to act now. Conservation Science and Practice, 1(8), e80.

- *Clean up your food*: They love our food almost as much as we do, so be sure to clean up so that you don't get them all hanging out under your stove.
- *Be sure that your screens and the space between the floorboards is blocked*: They can get through tiny cracks so be sure the space between the wall runner and the floor is closed and that there are no gaps in the window screens. It should reduce the number of insects who randomly stumble in.
- *Vacuum regularly*: Many love hair and skin flakes, so vacuum up all that pet fuzz so that they don't eat it.
- *If you do see a bug, try to figure out what it is before killing it*: Chances are what you see is harmless, but don't immediately fear it. Approach it with curiosity and respect.

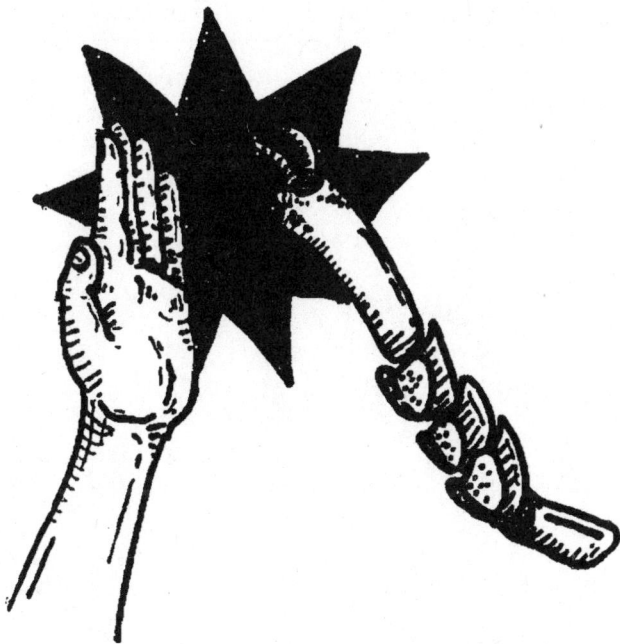

As my friend Dezi has often said, teamwork makes the dreamwork

Conclusion: How Do We Feel About Them Now?

*A*s the last twenty-one chapters have hopefully shown, bugs in all of their forms are the foundation of our ecosystem. Even those we find annoying are some of our most important decomposers (like cockroaches). They have been crawling on this earth for at least 500 million years longer than our oldest ancestors. Without them, the planet would collapse. A large part of our discomfort with bugs comes from a lack of knowledge and a separation from them. If the only insects you ever see are eating things in your home, it makes sense that you would generalize. In order to appreciate or tolerate them, we need to know more. We need to understand why they do what they do. We need to understand that they are just doing what they've done for hundreds of millions of years. Separating ourselves from them does not do us any favors. I hope that this book has been a stepping stone, for you to want to learn more and begin to appreciate what may have once been mystifying or scary. Neither of us could survive without the other, so we need to learn how to thrive together.

FURTHER READING

Dirzo, R., Young, H. S., Galetti, M., Ceballos, G., Isaac, N. J., & Collen, B. (2014). Defaunation in the Anthropocene. *Science*, 345(6195), 401–406.

Forister, M. L., Pelton, E. M., & Black, S. H. (2019). Declines in insect abundance and diversity: We know enough to act now. *Conservation Science and Practice*, 1(8), e80.

Fukano, Y., & Soga, M. (2021). Why do so many modern people hate insects? The urbanization–disgust hypothesis. *Science of the Total Environment*, 777, 146229.

Hallmann, C. A., Sorg, M., Jongejans, E., Siepel, H., Hofland, N., Schwan, H., ... & De Kroon, H. (2017). More than 75 percent decline over 27 years in total flying insect biomass in protected areas. *PloS one*, 12(10), e0185809

Isbell, F., Balvanera, P., Mori, A. S., He, J. S., Bullock, J. M., Regmi, G. R., ... & Palmer, M. S. (2023). Expert perspectives on global biodiversity loss and its drivers and impacts on people. *Frontiers in Ecology and the Environment*, 21(2), 94–103.

Janicki, J., et al. (2022, Dec 6). *The collapse of insects*. Reuters Graphics. reuters.com/graphics/GLOBAL-ENVIRONMENT/INSECT-APOCALYPSE/egpbykdxjvq/

Mcafee, A. (2020, Nov 4). The problem with honey bees. *Scientific American*. scientificamerican.com/article/the-problem-with-honey-bees/

Peterson, C. (2023, Oct 8). 'Like butter for bears': the grizzlies who dine on 40,000 moths a day. The Guardian. theguardian.com/environment/2023/oct/08/grizzly-bears-eat-moths-wyoming

Resh, V. H., & Cardé, R. T. (Eds.). (2009). *Encyclopedia of insects*. Academic press.

Ricci F. (2012). Social implications of malaria and their relationships with poverty. *Mediterranean journal of hematology and infectious diseases*, 4(1), e2012048. doi.org/10.4084/MJHID.2012.048

Thomas, C., Jones, T. H., & Hartley, S. E. (2019). "Insectageddon": A call for more robust data and rigorous analyses. *Global change biology*.

Xerces Society. (n.d.). *Understanding Neonicotinoids*. xerces.org/pesticides/understanding-neonicotinoids

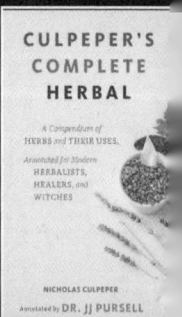